ORDER IN NATURE
VERSUS THE
ORDER IN PHYSICAL THEORY:

NEW METHODOLOGY
<u>IN PHYSICS</u>

Leah W. Ratner

Other books by Leah W. Ratner include

Non-linear Theory of Elasticity and Optimal Design:
How to build safe economical machines and structures;
How to build proven reliable physical theory

ORDER IN NATURE VERSUS THE ORDER IN PHYSICAL THEORY:

NEW METHODOLOGY IN PHYSICS

Leah W. Ratner

In memory of my parents Wolf and Fanya Marinberg and my sister Dina Marinberg

Subjects: Philosophy, Physics, General Mathematics, Logic, Classical Mechanics, Theory of Elasticity and General Theory of Change

Preface

"The lack of a conceptual framework suitable for the comprehension of physical reality apart from the act of observation has been elevated to the status of epistemological dogma by the overwhelming majority of contemporary theoretical physicists." (Gravitation, Electromagnetism and Quantized Charge: the Einstein Insight, by Daniel M. Pisello)

Since the time physicists have turned their attention to the studies of elemental particles incompatibility of micro-processes with the gravitational law of classical mechanics has become apparent. But, physics as a unified science is impossible if the general laws depend on the objects of investigation. Physicists see the possible solution of this problem in the unification of classical and quantum mechanics. For unification some background work in analysis of methodologies in physics has to be done. The current state of affairs in physics characterizes the lack of a common foundation. There is no common philosophical platform in physics; there are no logical rules for building theories; there is no connection between abstract mathematics and the concrete demands of physics; there is need in reliable criteria for testing correspondence of a physical theory to physical reality. "Physics has evolved and continues to evolve without any single strategy." (Encyclopedia Britannica)

Modern physics is divided into two branches: theoretical physics and experimental physics. But, the logical connection between these branches has not been established. And without such a connection it is impossible to prove physical theory. This book describes the common methodology that combines means of logic, mathematics and experimentation for building and proving physical theories. The book is not intended to review all concrete theories in physics from the point of view of the new methodology. However, this new methodology is the reliable way for reconstructing physical theories in order to have a common comprehensive ground in physics. The first theory, which is built on this foundation, is my new theory of elasticity. The book, *Non-linear Theory of Elasticity and Optimal Design: How to build safe economical machines and structures; How to build proven reliable physical theory,* was published in 2003 by Elsevier/Physics and Astronomy. This book, *Order in Nature versus Order in Physical Theory*, is an outgrowth from my previous book. The lack of a common philosophical platform in physics and the lack of a common methodology for building and proving physical theories is the obvious disadvantage for further development of physical sciences and technology.

Physics is not only an empirical science but theoretical as well. Ideas in physics are commonly presented as theories in mathematical forms. Physical theories are, indeed, our inventions. Their correspondence to reality has to be proven. However, physics has no methodology for proving theories. Physics, according to the Encyclopedia Britannica, is an experimental science that has no scientific methodology supporting it. This unique work provides analysis of the proposed logical structure of a physical theory and the criteria of proof for such a structure.

Since the beginning of the 20th century, new ideas have been emerging in physics. These ideas in the theories of relativity and quantum mechanics shook the foundation of classical mechanics. Then, works that criticize the basic laws of classical mechanics appeared. However, this criticism remains subjective until the common demands to prove the theories are formulated. While neglecting the previous achievements in physics, the critics also don't offer consistent new theories that can be tested and proven logically and empirically. A theory can be proven if the logical/mathematical structure of the theory is consistent, if the selected terms fit the logical structure of a theory and if such a logical/mathematical structure is supported by the facts.

This book contains the following sections: *Whole-Parts*; *Cause-Effect; Logic* in *Physics*; and *Metaphysics in Physics*. These sections discuss the types of relations that can be described within a physical theory. The parts, *Experiment in Physics, Mathematics in Physics and Logic in Physics*, analyze methods suitable for building sound physical theories. The discussions of some physical hypotheses, laws and possible new models of the physical universe can be found in the chapters, *Laws of Classical Mechanics, Theory of Elasticity, Light* and *Quantum Mechanics*. The chapter, *Physics* and *Physicist*, considers possible connections between subjective and objective elements in the processes of the studies of structures in the universe and construction of theories. *The Theory of Change* is a description of the common principles in structures of physical theories. The summary of main principles in physics is in the chapter, *Hypotheses and Facts*. *My Journey to New Methodology* reflects my own engineering experience. As a whole this book is liberating a reader from dogmas in physics and gives a comprehensive method of research. *Graphical Illustration of the Theory of Change* is given at the end of the book.

Note: The epigraphs to the chapters may support or oppose my points of view in the book.

Acknowledgements

I must thank first of all Francesca Ann Ratner who copyedited this book.

I am thankful also to my circle of support: my sons Jacob and Ilya, daughters–in-law Lucy and Christy; my grandchildren: Francesca and Misha; my brothers Simon and Boris Marinberg and sister–in–law Linn; my cousins: Solomon Davydov; Mark and Rita Davidov and Bassa. Marinberg; my friends: Sulama and Gene Resnick; Luda Babchinitser, and Traudi Unden

Contents

Introduction

"Physics deals with the most general properties of matter, such as behavior of bodies under the influence of forces. In the discussion of this question, the mass and shape of a body are the only properties that play a significant role, origins of those forces and the composition of matter often being irrelevant." (Encyclopedia Britannica)

An outline of the subject matter of the particular science is important for understanding the substance of science. The above definition of physics in the Encyclopedia Britannica refers only to the part of endeavors in physics that are concerned with building physical theories. Subject matter of physics includes studies of objects and their properties; studies of the processes making changes related to the objects; and studies of the forces acting in the universe. The objects participating in mechanical interactions, though, may change their geometry and/or relative positions, but they essentially remain the same integral wholes. The main purpose of a physical theory is to describe change of a physical whole or change of its position while the external unbalanced force is acting on the whole. In physical theory we are establishing exclusive quantitative relations among the change of a whole, resistance of a whole to the change and a force that is needed to induce the particular change.

In physics we also study mechanical and electromagnetic properties of the physical whole and the properties of the parts constituting a whole. Those parts have properties that under the certain conditions may demonstrate themselves as the properties of a whole. It is necessary to distinguish the properties of a whole from the properties of its parts in order to make a reliable analysis of change. For instance, according to the new theory of elasticity the limiting stress of a structure that demonstrates itself in the experiment is the relative value of the different origins. It can be the limiting stress of the material, depending on the properties of macrostructure of material, or it can be the limiting stress depending on the geometry of a whole structure. Both limiting properties should be taken into consideration for establishing the limiting stress of the structure.

The understanding of "whole" becomes more complex in the 20th century with the discoveries of structural parts of a whole, and ultimately the discovering of elemental particles constituting the whole. It appears that a whole is a hierarchy of parts situated at the different levels of structural division. In the contemporary view, not only matter consists of elemental particles and structures, but energy also can be divided into smaller units of energy.

Matter and energy exist in the structured space of a whole. The space of a whole has its characteristics. The structure of any whole depends on the distribution of energy in the space of the whole. Physicists create hypothetical models of wholes in order to understand their properties and behavior better. It is possible to study the properties of a whole complex body or properties of a part. The relationships between the properties of a geometrical whole and the properties of its parts can be revealed in the processes of change rather than in a state of equilibrium. Dynamics of present predicts future.

At this stage of development in physics we do not solely deal with the philosophy, laws and methods of classical physics. New science that appears at the beginning of the

20th century, namely quantum mechanics, does not operate within the laws of classical mechanics.

> "Attempts to understand the motion of the electrons going around the nucleus by using mechanical laws – analogous to the way Newton used the laws of motion to figure out how the earth went around the sun – were a real failure..." (QED, Richard P. Feynman)

Classical mechanics, in turn, have not yet adapted the, statistical in nature, principles of quantum mechanics. The reason is that classical mechanics has more definitive and comprehensive methods for providing explanations for the dynamical events in mechanics. If we assume that the laws of physics depend on the objects of study, then the question arises: Do we deal with the same physics? There is no doubt that in order to be unified physics should have uniformity of methodology; i.e. uniformity of its view on the structures of objects small and large; uniformity of its laws; and uniformity in the methodology of the creation of theories. One may consider that the current explanations we have of the physical sciences is a crisis in physics. Physics at this stage deals with the particulars while having no common philosophy and logic. Thus, the starting point of a discussion in this book is clarification of the scope of the fundamental problems in physics.

The studies of elemental particles and their structures lead to the division of physics in areas having different approaches of inquiring knowledge. There are two main approaches in physics that are useful – the empirical studies of properties of the objects existing in the universe and the creation of the common logical-mathematical structure for the theories describing the different physical processes. Methodologies of those endeavors in physics are quite distinct. Our models of the natural structures in the universe are hypothetical; we cannot prove them. This is because the models of nature have no logical structures, and nature itself doesn't have logic. On the other hand, we can construct a logical-physical theory of change and prove or disprove it. Methodology of building and proving physical theory depends on the logical structure of a theory and the rules for selecting the physical terms fitting this logical structure.

There is an order in nature. Physical knowledge wouldn't be possible if we only saw chaos and not the basic order in the universe. Chaos also has a place in nature as an essential part of the organization of the universe, for in order to organize the structures it is necessary to have the bulk of disorganized matter and space for their organization. But, "order" does not mean "logical order." It can be order depending on the structure of the universe. In physics we are concerned with the hierarchy of parts of the universe. The studies of structures of the universe are only a part of what physicists do in physics.

The universe is a dynamic place. A physicist is also interested in the physical processes and in the prediction of the possible consequences of those processes. An engineer, in turn, is interested in the application of specific physical knowledge in technology. The knowledge of the processes should be reliable for a safe and economical application. The processes in physical theories are described in the form of mathematical functions. At the same time we should not assume that knowledge of the structures and the forces in the universe can be translated into mathematical descriptions of the processes. In a theory we combine the empirical knowledge about an object, a logical structure of the theory and the mathematical rules of operation.

One of the most important questions that we ask in epistemology is: Does a physical theory reflect the order in the universe, or does a theory have a different nature, contents and laws than those found in the universe? Currently, a physical theory is not differentiated enough, if at all, from the studies of the universe. However, for developing the successful methodology of construction and proving physical theories we have to distinguish order in nature from the order in physical theory.

There are laws of nature that make the universe a place of order. We can discover some of those laws. But, in general, the laws of physics that we use are our assumptions which we need for outlining the boundaries of our research in physics. It is possible that some laws of classical mechanics need to be corrected. The universal laws are necessary for building theories. Empirical support of the universal laws is required for all particular cases. If there is no universal support for some fundamental assumption-law, then the law needs revision. In order to have a common ground for physical studies, our models of the physical world and physical theories should comply with the accepted universal physical laws, such as the law of conservation of mass-energy, the law of conservation of momentum of a closed system, the law of elasticity and the other particular laws which correspond to the law of conservation of energy.

In order to differentiate between the studies of objects and their structures in nature from the logical structure and concepts of a physical theory it is necessary to analyze commonalties and differences of those studies. The theories are not reflections of the structures and the forces found in nature, and never were, as it has been properly stated in the Encyclopedia Britannica. The objects and the processes in nature have no known purpose to us. Our own actions and inventions have purpose. The main purpose of a physical theory is to describe a physical process or a phenomenon in a way that allows the predictions of the process to have development and limitation. The studies of the structure of matter and characteristics of its parts are mainly empirical. Studies are successful if the facts about the structure of matter and the characteristics of parts of space-matter are obtained. The creation of a physical theory, on the other hand, is an analytical and experimental process. It is successful if a theory has provable logic structure and is supported by the facts.

Besides his hypothetical ideas and observations for the creation of a theory, a physicist needs to implement methods of science, which are not given to us by nature, such as logic, mathematics, physical laws and definitions. We should not ascribe our methods to nature, for it is rather a metaphysical idea that if we successfully use those methods then we have discovered them as hidden in nature. Our methods are convenient tools which we have invented for our own practical purposes.

For building a physical theory we need the consistent logical structure of a theory. It is worth mentioning, that until recently, no attention has been paid to developing such a structure. The result is that physicists and philosophers of science accept, and even embrace, the defensive idea that a theory can be disproved but it cannot be proven. An introduction of a logic scheme as a requirement for the provable structure of a physical theory changes that discouraging nihilistic approach. A physicist has to select the characteristics of an object in such a way that fits the conceptual logical scheme. The scheme does not pertain to the hierarchy of matter and forces between particles, rather it is concerned the connection between measurable change of the object under investigation with the selected empirical statistical characteristics of a whole, empirical terms and the theoretical terms related to the change. The theoretical terms are selected based on a prior knowledge of physics.

The method of validating a theory differs from the empirical methods of testing the properties of structures in the universe. The facts do not need logical justification. Nature has no known purposes for us and no need for justification. On the other hand, our hypotheses about the relations of selected characteristics of a whole have a purpose and need for logical justification. That can be achieved only by introducing the objective logical structure as the basis of a theory. In physical theory, besides subjective ideas, we also use objective methods of logic, mathematics, experiments and our prior knowledge. The physical world that is described in a physical theory doesn't follow from the knowledge of elemental particles and their structures, but does not contradict it, either. In a physical theory we select physical characteristics and terms to fit a logical scheme, however, we select real physical characteristics rather than our "inventions" that we ascribe to nature.

Usually, the properties of a body/whole are identified with the properties of its matter. In fact, what is missing in such a presentation is that matter occupies some space. A whole has geometrical form and inner structure, or even a hierarchy of organized space-matter structures which affect its properties. There is no matter without space. When we speak about the properties of a whole we speak about the properties of particular space-matter.

In a physical theory we are analyzing the properties and changes, or possible changes, of a whole which depend on the geometry of the whole. The properties of elemental particles, such as the strength of a nucleus, strength of an atom or strength of an atomic-molecular structure cannot be used for determining the strength of a whole. The properties at different dimensional and energy levels are different. Each level presents its own world. We cannot compare them mathematically and physically. Each world has its own units which are different from the units of space and time at a different dimensional level. Nature achieves balance and stability at each level independently. When balance is disturbed for a whole structure the structures in the foundation of the hierarchy of space-matter usually survive. On the other hand, the distraction of elemental basic structures leads to the distraction of a whole. Knowledge of the elemental building blocks of matter doesn't guide us to the complete knowledge of an object, even though the object may consist of elemental particles and their structures.

For all practical purposes, the methodology of building a physical theory needs to be separated from the studies of the elemental building blocks and structures in nature. In practice those studies are separated, but they are not separated in the philosophy of physical science. Our philosophical ideas about the physical world may present support, as well as obstacles, to inquiring knowledge in physics. Physical theories describe mechanical interactions of independent entities. Such interactions, within limits, do not change the essence of the entities participating in an interaction. A physical theory considers the changes of an entity that are induced by outside forces. The main purpose of a theory is to describe the change of an object in order to predict its behavior through the changes of the selected characteristics.

In building a theory we employ subjective ideas, as well as objective scientific methods, such as logic, mathematics, a prior knowledge of physics (namely, physical laws), geometrical theorems and empirical methods. These methods allow us to describe, test and correct, if necessary, any subjective ideas. For a description of a phenomenon or process we select the concepts that are connected to change. The logical-mathematical structure of a theory consists of the propositional equation describing the possible connections between selected variable concepts and the inferential equation for considering the rate of change. The rate of change is a good indicator of the possible behavior of an object. The test of a theory consists of testing the system of propositional and inferential equations with their correspondence to the logic scheme, a prior knowledge/laws and experimental data. We can prove physical theories because we construct them according to some logic scheme and rules.

Unlike our ability to prove theories, we cannot prove our models of the universe and explain with certainty the relations between its parts. All our explanations are, at best, plausible. The reason for this is that we deal with fragments of the different worlds at different energy levels, which exist separately from each other. The relations between the different independent entities cannot be described with a logic structure. A certain provable relationship can only be established between the inseparable characteristics of an entity, such as the geometry of an entity and its matter. Change of the space occupied by a whole leads to the change of energy of a whole. On the other hand, the relations between a whole and its parts can be established only as statistical empirical data. Functional relations in physics do not exist between the independent entities. A whole can be physically divided into its independently existing parts. But, we cannot establish a provable functional relation between the independent parts. This point of view has logical implications. One of the implications, for instance, is that there is no functional connection between cause and effect, for cause and effect physically belong to different independent entities.

Whole – Parts Relationship

"By and large, a system in classical physics can be analyzed into parts, whose states and properties determine those of the whole they compose." (Stanford Encyclopedia of Philosophy, Holism and Non separability in Physics)

The point of view in this book is that the properties of a whole are not the properties of its parts. There is no functional relation between the properties of a whole and the properties of parts of a whole. Experiments clearly continue to demonstrate this position. The properties of a whole can be found experimentally. A theory, on the other hand, describes the relationships between change of a whole and the selected properties of a whole.

Our studies in physics refer to a physical entity or, in other words a whole or an object. The properties and the relative position of a whole are the starting point of investigation. The question addressing what is changed in a process and how, would be impossible to answer without the concept of a "whole," which existed before and after a change took place. A whole can be an elementary particle, the nucleus of an atom, an atom, an atomic-molecular structure, the macro-structure of a material, a common body, and a system of common bodies, a galactic system or a formation of electro-magnetic energy, such as a beam of light. The concept "whole" refers to organized space-matter.

The common features of the objects governed under the concept "whole" are as follows: A whole has geometrical form, surface and mass; a whole occupies space and consists of its parts. When we speak of the properties of a whole we are referring to the properties of concrete organized space-matter. A whole may have a hierarchy of parts which are arranged at different levels of inner energy. The law of conservation of energy in a closed system applies to such a whole. The structure of a whole affects its properties. A part of a whole may also consist of parts, and can be viewed as a whole with all the common properties of a whole. The specific properties of a geometrical whole are not the properties of the parts at any level in the hierarchy of matter. It is necessary to study the geometrical and material properties of a whole. The studies of elemental particles and elemental structures do not yet lead to the knowledge of a whole, which consists of its particles and structures.

In different experiments we may observe characteristics that can be properties of whole or statistical properties of the parts. It is necessary to distinguish between them. In order to make such a distinction we need a conceptual understanding of the whole-part relationship. At this stage of modeling a whole in physics it is not difficult to imagine a whole, of any form, as a container which contains smaller objects of the same or different materials. A subdivision into smaller and smaller parts may continue. The properties of elemental structures are not that important for obtaining the properties of a whole, such as the resistance of a whole to change and its limiting strength. Those characteristics of a whole depend on the geometry of a whole and the statistical experimental characteristics of the material, rather than on the characteristics of the smallest parts. Such parts not only have different incomparable dimensions but incomparable properties as well. The differentiation between the properties of a geometrical whole and the properties of the parts of a whole should be made. In order to make this distinction, whole and parts should be studied as separate entities. Though it is impossible to establish a functional connection between independent entities, it is possible to establish a functional connection among the characteristics of an entity. Change of one characteristic leads to the certain change of another characteristic. For example, a change of the geometrical form of a body caused by an external unbalanced force leads to the change of the inner energy of a body. This force demonstrates itself as a stress in the body.

Each level of structural division of a whole presents its own world. A balance of the forces is reached independently at each energy level. Knowledge of the properties of the elemental particles will not give us knowledge of the properties of the whole containing those particles. For considering the relations between a whole and its constituting parts; and the relations between the parts at the same level and the relations between a whole and other wholes – it is necessary to clarify a philosophical platform in science. The seemingly abstract metaphysical questions, such as a whole-part relationship, or the relation of the geometry of a whole to the forces acting in a whole, have immense importance for understanding physics and its practical applications. In their foundations physical theories have certain philosophical points of view, even if these points have not been accentuated.

There are different points of view on whole-part and whole-whole relations in contemporary physics. Classical mechanics views a whole as a sum of its infinitesimal parts. A whole can be divided into smaller and smaller parts down to its elementary particles and their structures. Ultimately, a whole is viewed as a sum of these parts. This view is extended onto the properties of a whole. This assumption would be no problem if we considered such properties of a whole as its mass. But, even volume of an object as a whole is more than the sum of the volumes of its parts. A whole is not the sum of its parts. This point of view is especially important when applied to the knowledge of the properties of a whole. For example, a property of an elemental structure such as its strength does not represent the strength of a whole structure. The strength of an atom is not equivalent to the strength of its nucleus. The strength of a molecular structure is not the strength of the atoms building the molecular structure. Furthermore, knowledge of the properties of the parts of a whole cannot be used for obtaining similar characteristics of the whole. A whole and any of its part are different entities that have different characteristics. The characteristics are often not comparable.

There is the idea in physics that between particles and parts of a whole there are forces acting, keeping the parts of a whole together. In his Theory of Relativity, Einstein came to the conclusion that ether, as a medium, is not needed. In general, this is to say that there is no need for gluing the universe together with some medium-ether. Currently, there is a hypothesis that considers some particle forces as the means responsible for keeping matter of a whole together as an entity. This hypothesis, as well as the hypothesis of ether, is not sustainable. The particles, which can be active or neutral, exist in a whole, but they do not have the exclusive role of keeping an entity stable. The point of view here is that if a whole is not in the process of change, then no internal forces are required to keep a whole as it is. Only outside force or energy acting on a whole may change its geometry and induce stress, which may not only change but can destroy the entity, as well.

In physics there is no clear distinction between the concepts of an "object" and a "phenomenon". Some physicists insist that physics does not deal with the objects in the universe, rather with the processes. In the book, *Three Roads to Quantum Gravity*, Lee Smolin wrote, "The universe is made of processes, not things." In the process of analysis in physics we can and should distinguish a process from an object. A process is the change of an object while an unbalanced external force is acting upon a structure that we call "whole." A process may continue for a long or very short period of time, but in all cases, process refers to the change of an object and its energy. A physical object, on the other hand, is an isolated integral physical system to which the laws of conservation of mass-energy and conservation of momentum are applied.

If a whole consists of independent parts, then the question arises: What is keeping these parts together as a distinctive whole? In regard to this matter, there are number of hypotheses in physics. There are field theories, such as the gravitational field and electro-magnetic field theories. Gravitational field depends on the mass of the bodies in the field. It is assumed that every object in the gravitational field is subjected to a gravitational force, which is the result of interaction between objects situated at a distance from each other. On the other hand, action occurring at a distance contradicts the basic idea in physics, namely, that there is no action at a distance. Consequently, the idea of the four universal forces that keep objects and the system of objects in their natural forms and places developed in modern physics.

"One way to understand the new unified theories is to approach them by way of the four fundamental forces of nature – gravity, electromagnetism, and the "week" and "strong" nuclear forces." (The world treasury of Physics, Astronomy, and Mathematics, Timothy Ferris)

Celestial systems and all common objects are under the influence of gravitational forces. "Gravitational force" is assigned to the elemental particle called "graviton." Gravitons have not yet been discovered experimentally. Electro-magnetic force is generated around positively or negatively charged objects. This force exists between electrons. Strong and weak elemental particle-forces act inside the nucleus of an atom.

Einstein proposed the hypothesis of space geometry that controls the positions of the parts of a system.

"Space-curvature is a purely physical characteristic, which we may find in a region by suitable experiments and measurements, just as we may find a magnetic field." (Spherical Space, Arthur Stanley Eddington)

String theory assumes that connections are realized through strings, which are energy structures in ten-dimensional mathematical spaces. Any hypothesis can be described mathematically. However, mathematics doesn't provide proof for the hypotheses. None of these models have been proven.

The idea in this work is that integration of a physical entity is reached, mainly, with the distribution of mass-energy in space. There is no unbalanced force can be detected in a balanced whole. Force in a whole appears only when an external unbalanced force is acting on the whole. Although none of the hypotheses or models of structures in the universe can be proven, a model, at least, should be in conformity with the universal law of conservation of energy and momentum.

Summary: A whole is an object that has geometry, surface and mass. It occupies space and consists of parts or a hierarchy of parts. In turn, a whole is also a part of a larger system of wholes. Classical mechanics and quantum mechanics study wholes which are situated at different dimensional and energy levels. The theories of classical mechanics describe interactions of bodies coming in contact. Those interactions are considered to be at the same dimensional and energy levels. As a result of interaction, changes in the bodies occur. The changes of the bodies that are in contact are different even though the forces acting on the bodies have the same values. This occurs because the bodies in contact are different. Change in one body is not functionally connected with change in another body. There is no functional relation that can be established between independent integral wholes. This is likewise true for the relationships between a whole and the parts constituting a whole. The parts of a whole exist at different dimensional and energy levels. The properties of a whole depend on the geometry of space of the whole. In the course of mechanical interactions an exchange of energy occurs. The properties of the wholes participating in the mechanical interactions are not transferable. Knowledge of the mechanical properties of the parts constituting a whole doesn't lead to the knowledge of the properties of a whole. Whole consists of hierarchy of parts in equilibrium. Whole could not exist if it is depending on internal forces to keep its parts together. Such model depending on forces and thus on constant changes contradicts the law of conservation of energy. We cannot prove our models of structures because functional relations can be established only among physically inseparable characteristics of a whole. In order to analyze the change of a whole we consider the cause–effect relationship.

Cause-Effect Relationship

"[N]o event comes to us labeled "cause" or "effect".
An event has to be deliberately *taken* to be cause or
effect. Such taking would be purely arbitrary if
there were not a particular and differential problem
to be solved." (Logic, John Dewey)

A common way of reasoning in physics is to assert that
if some change occurs then there should be a cause for it.
In logical terms, change of an object is the effect. Cause
and effect are useful logical concepts in the analysis of
events in physics. Therefore, one of the most important
tasks of epistemology in physics is to explain the cause-
effect relationship. Cause-effect relationship results from
conjunction of at least two independent objects. When
objects are in conjunction, and an exchange of energy
occurs between the objects, this becomes a starting point of
a physical event. For understanding, and descriptive
purposes the role of "cause" is assigned to one of the
objects and another object demonstrates "effect." Change
of an object in logical terms is effect. There is no functional
relationship that can be established between cause and
effect in physics. This is due to the fact that cause and
effect belong to different physical entities.

Do cause and/or effect have some special features that
make them different? There is no inherent physical feature
that makes an object in conjunction become a cause or
effect. The roles that we assign to objects in contact
depend on the purpose of an investigation. Natural
phenomenon does not possess a purpose nor mind of its
own, and no roles of cause and effect that are assigned by
nature. Though, a condition may arise or we may create a

condition with a purpose in mind to use some object as a cause in order to investigate another object. For example, we can be interested in studying the waves formed from a boat crossing the lake. Then the boat is a cause of the waves. Or, in the case when we are interested in the stability of the boat, then the boat is an object of study. In logical terms, we study an effect and the waves are the cause of the possible instability of the boat. Analysis of a conjunction shows that we may describe an event as the result of an action of two forces created with equal values and having opposite directions (Newton's Third Law of Motion). Each force is the cause of change of the object to which it is applied. Though the forces are equal, the objects are usually different, and the consequences, or effects, of contact are different. We study the effect in the object of interest.

What kind of relations exists between cause and effect if there is no inherent feature that distinguishes cause from effect? There is no functional relation that can be established between cause and effect, as functional relations can only exist amongst the properties of an integral whole rather than between independent entities. In logical analysis, if a certain effect is of interest to us in one entity, then the other entity is a cause that transmits its energy to the entity-effect. Energy is transmitted on contact. Which one of the objects in contact appears first in the world is not necessary for describing change. When an event takes place, both objects are in the same place at the same time. There is no need in physics for building a chain of events which leads to the initial contact if we are interested in the changes and behavior of a particular object. For the description of change of an entity we consider the concrete occurrence for which "cause" is "force".

Change of an object, or its relative position in space, can be presented with a function. There are certain demands for such a function. In order to describe a limit of the relations a function should be non-linear. A linear function can not describe a process or a change that has a limit. The complete description of a non-linear function in a plane includes a system of two equations, namely, the basic conceptual equation and its derivative that describes the rate of change of the function. Only the consistent logical system of the equations may give a definitive description of a change. This mathematical system needs to be tested experimentally and logically. The function of change, if related to a family of similar objects, may become a proven predictive theory. It is impossible to predict the behavior of an individual object from observation. However, a prediction is possible with a theory of comparison of the changes for the members of a family of objects.

A propositional equation in physics is not a description of the cause–effect relationship. It is only a description of the effect. The equation of change connects the change of an object with the force that is required for the particular change and the resistance of an object to the change. All concepts in the function of change belong to the object in which the effect takes place.

The relationship between cause and effect in physics is presented only in an equivalence of energy induced by a cause and realized in an effect. The roles of cause and effect are exchangeable. The behavior of any independent entity can be studied independently from the cause. Thus, change depends on the force acting in an entity and on the resistance of an entity to the particular change. The function of change and its derivative function allow us to predict the possible behavior for each member of a family.

Note: There are internal forces of decay of whole. Those forces can be considered as a cause of change.

Summary: The concept "cause" refers to an external force/ energy that is applied to an object causing changes in the object. Physically, cause refers to one of the bodies in contact. The other object in contact in logical terms is an effect. Change of the certain characteristics of an object-effect can be described with a non-linear function. Change of an object, or change of its position, is proportional to the force, and inverse to the resistance of a body to change. The theory describes the predictable values not only for a single event or individual object, but to a family of similar objects. A general theory cannot exist if no comparison of changes in different objects is made. For the construction of a theory we have to separate a cause from an effect. Such analysis allows us to adequately understand and describe a physical event. Physical relations also should be alienated from the metaphysical statement

Metaphysics in Physics

"We may accordingly define a metaphysical sentence as a sentence which purports to express a genuine proposition, but does, in fact, express neither a tautology nor an empirical hypothesis. And as tautology and empirical hypotheses form the entire class of significant propositions, we are justified in concluding that all metaphysical assertions are nonsensical." (Language, Truth and Logic, A. Ayer)

"But it is well to bear in mind that God does nothing out of order. Therefore, that which passes for extra ordinary is so only with regard to a particular order established among the created things, for as regards to the universal order, everything conforms to it." (Discourse on Metaphysics, Gottfried Wilhelm Leibniz)

Physics is distinguished from metaphysics, first of all, by its methodology. Methodology in physics should lead to concrete provable knowledge. Metaphysics, on the other hand, is not concerned with such proofs; it is rather concerned with the pure idea of a predetermined causation of events. However, metaphysical ideas and researches appear in physics for we have no clear methodology to separate metaphysical ideas from physical. There are several principles should be kept in mind in regard to this task. Our models of space – matter structures cannot be proven logically. Validation criterion for such a structure is the general law of conservation of mass-energy and momentum. An idea to establish a functional connection leading from an event to "first cause" is a metaphysical idea would "first cause" to be an idea, program, or mass-energy. The idea to connect logically an event to some cause separated in space and time is the metaphysical idea. On the other hand, functional connection can be established between change of an entity and the related characteristics of entity. Dynamics of present allows predict future. The first question of epistemology to ask: Is it possible to achieve an objective knowledge of the physical world? The answer to this question depends on the philosophical position. Science, physics in particular, gives a positive answer based on the possibility of obtaining adequate knowledge. Optimism in science is built partly on experience, and partly on belief. We believe that because we are successfully learning the matter-structures and forces in the universe this knowledge may give us a complete understanding of nature. There is the persistent idea in physics that knowledge of elemental particles and universal forces may lead to knowledge of all objects and processes in the universe.

The point of view here is that the properties of a whole are not the properties of its parts and cannot be deduced from the knowledge of elemental particles and universal forces. The reason for this is that in a hierarchy of structures we deal with independent entities at different energy levels in space. The properties of the parts at one level are not transferable to the next level. We cannot infer the properties of a whole based on the properties of parts constituting a whole. There are no necessary relationships between independent entities that can be described with a function.

The same is true for cause-effect relationship. Cause is represented by one entity, and effect is represented by another. History shows that we may reach a progress in technology, even with incomplete or distorted knowledge. This brings optimism in science. But for true, steady, cost-efficient progress in science it is better to have more organized and adequate ideas than are currently exist in the methodologically disconnected parts of physics.

Metaphysics answers in the negative on the question of whether or not there is a possibility of understanding nature using the scientific methods of physics. In this view not everything is open to physical investigation. Thus, the cause of a cause of creation of the universe, or first cause, is unknown and cannot be known. This conclusion in metaphysics is based on the logical position that everything, including all things and processes, has a cause. Then there must be a cause preceding that cause and so on. Determinism in knowledge is possible only if the first cause is known. We never will be able to investigate the first cause and understand it as a purely physical occurrence. .

One of the differences between physical and metaphysical methodology is that in physics the investigation practically starts and proceeds from interrupting the infinite chain of causation for a single event or phenomenon. Metaphysics, on the other hand, approaches the question of understanding a phenomenon from a logical determinism point of view. Logical determinism is a philosophical point of view of the event that is casually determined by an unbroken chain of events that stretches indefinitely back in time. If there is even one indeterminate event in this chain, then no determinism can exist in physics. For example, principle uncertainty in quantum mechanics, seemingly, abolishes determinism in its classical definition. However, in quantum mechanics there is a way to predict quantum events with the same accuracy as calculating the deformation for an elastic body. Erroneously, philosophical determinism from the initial cause all the way to the effect, doesn't lead to any physical predictions, while the method of isolation of an object, in which event takes place, does make the reliable predictions. Note that although determinism and accuracy of predictions in a physical theory are, seemingly related, in essence they have the different meanings. Accuracy depends on methods of measurement or calculations. Determinism refers to our point of view or philosophy of relations.

Physical determinism can be referred to a concrete event or phenomenon. But, if a physical effect is separated in time from a cause, then there is no physical certainty that physical event even occurs due to such cause. This is because a physical change occurs on contact and cause and effect appear simultaneously. In physics the inconsistency of scientific methodology is that philosophically physics recognizes causation in time, i.e. first comes causation followed by effect. But in practice we deal with selecting two objects, one object as a cause and another as an effect which appeared simultaneously in conjunction.

Contemporary science considers evolving matter-energy (creation of matter out of energy and vice-versa) as the first cause of the creation of the universe. The conclusion from such a position is that the first cause of everything is energy and elemental particles. Every singular object and process, carry on this matter-energy as the immortal elements coming from the beginning of creation of the universe. But, the properties of common objects are not yet proven as solely depending on basic structures and universal forces. The contemporary view on cause and effect in science also stops short of the principle of indeterminacy in quantum mechanics. Physical science is not yet truly consistent in its philosophical views.

By considering the relations between the geometry of a whole, forces and the statistical characteristics of matter constituting the whole a physical theory can predict the possible changes of a whole. This happens because we analyze physical unity of space and matter. The properties of an entity depend on each other in such a way that change of one property leads to the corresponding change of other properties. Although separation of matter from its form seems artificial and metaphysical, such a separation shows a real physical relationship. On the other hand, division of a whole into smaller and smaller parts is physically possible. Any of these parts can exist independently. However, no functional connection can be established between the independent entities. The idea of finding such a connection using knowledge of elemental particles and structures has no logical or experimental basis. Physics is distinguished from metaphysics by its methodology which ought to lead to provable physical knowledge. Metaphysics, on the other hand, is not concern with such proof; rather it is concerned with the pure idea of a predetermined causation.

Any physical theory deals with the accurate prediction of behavior and the limit of existence of an entity. A theory has to be supported with physical facts. But, the methods that lead to the proof of a theory are not solely empirical. In a theory we make logical-mathematical connections between the properties of a whole and therefore have to test the consistency of those connections. Logically necessary connections cannot be established between cause and effect as metaphysics implies. From the physical point of view there are no objects in the universe with physical properties that can differentiate them from being a cause or effect. Cause-effect separation is a very useful tool for analyzing a phenomenon. What we call cause-effect in physics appears simultaneously when independent objects come into contact. In the process of analysis we select an object that we will consider a cause, and another object that we will consider an effect. We assign the role of an effect to an object in which we study changes. The object that is of interest to us only from the point of view of the amount energy it transfers will be considered cause. We have scientific methods to study an effect even without employing knowledge of a cause because a cause is implemented in the effect. A prior accepted knowledge that we formulated as the universal law of conservation of energy allows us to attribute change to the energy released during contact. Thus, in a physical theory we describe the relations in an entity that we consider to be an "effect".

Cause-effect in its historical understanding, from an initial cause to an end result, bears not a physical, but rather, metaphysical character that is impossible to prove with physical methods. This approach is not constructive for building a physical theory. We are looking for a useful methodology. It is our metaphysical idea about causation that brings the notion of succession in the cause-effect relationship even where it does not exist. In some events, like in historical or social interactions, the intrusion of man's intentions, actions and manipulations makes cause and effect seemingly connected in time. For example, a man takes a bat and hits the intruder. What is the cause for the bone fracture in the intruder? In physical reality, rather than criminal, it is still the contact of a bat with a human bone that causes the fracture. Metaphysical element here is an introduction of human will and action as the cause.

In metaphysics the chain of cause-effect events proceeds from some first cause that has a purpose as it leads to the effect. For example, in metaphysics the idea of "intelligent design" is introduced as a cause that brings about the effect of humanity. In quantum physics it is the properties of elemental particles that have the ability to organize matter and in a chain of events build objects and the universe. The epistemological point of view is that first comes the cause followed by an effect is, indeed, metaphysical. The physical disconnection in time between a first or previous cause and a later effect leads to the idea of some special properties of cause that can be realized in the particular later effect. From the physical point of view there are no cause and effect roles assigned by nature. In the process of analysis of a phenomenon man ascribes to one part of a juncture of independent objects the role of cause-energy, and to another object the role of effect-change. Metaphysics comes into play as soon as physics has difficulties explaining a phenomenon. Our models of physical structures and explanations are, indeed, hypothetical. For an explanation of a model we consider its relatedness to universal laws. Nevertheless, our theories that connect change of a whole to its geometrical properties can be proven and we have methodology to do that.

Summary: The idea that first cause separated in time and space can be logically connected to present effect is a metaphysical idea. In metaphysics first cause is an idea/program; in physics first cause is mass/energy –both represent metaphysical causation. In reality there is no physical cause that exists in time before an effect takes place, for that would be a contradiction to the universal law of conservation of energy. An object cannot respond before a force that appears as a result of contact with another independent entity and changes the object or its position. Transmission of energy from one object to another occurs when objects come into contact. Only our metaphysical approach in the explanation of events brings the element of consequence, and thus the first cause into a physical analysis. The remote cause that allows metaphysics to appear in physics is excluded from a physical theory. Our explanations of physical events are still often metaphysical. We need to test a physical theory in order to exclude metaphysical explanations. The physical world consists of wholes and their parts existing at any different dimensional and energy levels. One may see it as one universe with many different independent worlds. That is how physical science actually treats them. The physicists who study subatomic particles consider them to be independent entities. A nucleus, atom and the atomic structure of an atom are all considered and studied as   independent wholes, rather than the parts of some larger wholes. Characteristics of a whole body have no effects on the characteristics of their atomic or subatomic particles and structures. Similarly, elemental particles and structures, though they are the structural parts of a body, have no effect on the mechanical properties of a body. Even of a small block of material, we do not calculate the mechanical properties using knowledge of the properties of particles. It is impossible to deduce the properties of a body from the properties of an atomic-molecular structure. We deal with different physical worlds that are incomparable, and separated by spaces dimensionally. The point of view that knowledge of elemental particles and universal forces may give us knowledge of all matter-space structures is a metaphysical idea. This idea cannot be tested with physical means. In order to test something in physics we have to compare the similar quantitative characteristics of objects. The physical characteristics are comparable at the same dimensional level. Physical theories consider properties that relate to each other and are comparable at the same dimensional level. It is the realm of physics.

The great achievements in physics of 20th century deal with the idea that space has properties and a shape. Space without matter is not an object of study in physics, but in combination with matter, space has its own characteristics. Physical theory is occupied with the studies of space-matter entities. A theory describes a functional connection between space and matter. This is possible due to an intimate connection between space of a whole and the matter occupying this space.

Experiment in Physics

"It is being increasingly recognized…that the world of experiment is not understandable without some examination of the purpose of physics and of the nature of its fundamental concepts." (The Logic of Modern Physics, Percy Bridgman)

A physical theory is our construction, rather than a work of nature. We need experimental tests for establishing the connection between our hypothesis and physical reality. A theory in physics can be presented with a logical-mathematical system of equations, namely, a propositional equation and an inferential equation. The experimental corroboration of a propositional equation doesn't yet mean that the physical hypothesis is correct. We can devise many variations of a propositional equation, which have the same experimental result. But only one of those equations can be physically correct. The experimental method is not the only objective criterion in physics.

The test of validity of a physical theory is also in the logical analysis of an inferential equation. For a given whole the physical relations are, indeed, exclusive. Besides the experimental verification of a result of a propositional equation we have to prove our selection of terms in the equation and the consistency of the logical structure of the theory. The experiment is one of the important tools in physics. From the point of view of epistemology an experiment ought to provide a connection between our hypothesis and physical reality. Another role of an experiment is testing and measuring the specific properties of an object as a whole and its parts. However, an experiment by itself is not sufficient for the verification of a theory.

Physical theory includes measurable and theoretical concepts/terms that cannot be measured directly. The theoretical concepts can be inferred from common definitions and universal laws. For example, such geometrical characteristic as the area of a cross-section, moment of inertia of a cross-section, geometrical stiffness, volume, wavelength, amplitude and alike can be obtained by measuring linear and/or angular dimensions of an object. The internal forces, on the other hand, which appear in the process of change of an object, cannot be measured directly or indirectly. We obtain their values through the mathematical-logical procedures from the common definitions and universal laws, such as the definition of work, the law of conservation of energy, Hooke's law of equivalence of stress and strain, etc. All theories should obey such universal laws. There is, however, the need not only in experimental confirmation of result of propositional equation, but testing representation of relationships in this equation. This can be proved by affirming logical-mathematical structure of a theory.

A common point of view is that the concepts of a theory correspond to the observable characteristics of a phenomenon, and therefore the satisfying experimental test of a proposition makes a theory acceptable until the facts to the contrary are discovered. Such an incomplete and inadequate methodology makes physical theory littered with doubt. Though the concepts of a physical theory have physical meaning, a selection of characteristics / concepts is subjective. In order to make the process of selecting the concepts more objective we have to apply some logical rules to a process. It is not enough to use common sense judgment. We select characteristics depending on the observations and logic of connections among the selected characteristics. Logical-mathematical analyses may give conclusions which are different from the conclusions that might be reached from an observation. The overall point of view on the nature of a physical theory helps to select the concepts of a propositional equation of a theory.

Physical theory describes a process of change in a whole object. The reason for this - it gives us the possibility to compare what was there before a change took place with the same characteristics after the change. In order to have certainty in our endeavors we need the law of conservation of energy and momentum in a closed system. Change means that energy was spent. It is a starting point. We view a whole as consisting of a hierarchy of its space-matter parts. However, in a theory we describe relations not among those parts, but among selected characteristics of a whole related to the change of a whole. The selection of concepts needs some formal methods of testing. Empirical methods cannot be solely responsible for selecting the concepts correctly. We create the theoretical description of a change, which only partly can be tested experimentally. Thus, the derivative equation cannot be tested experimentally for it is a logical-mathematical conclusion that has a relative value as its result. A relative value can be calculated but not measured. We need logical methods for testing such conclusions.

One way to construct an adequate physical theory is a proper selection of the properties-terms related to a change; the proper selection of a propositional mathematical equation and the proper logical-mathematical structure of a theory. However, physical theory is connected with the empirical data, but empirical methods are insufficient for proving a physical theory. A theory has to be presented in the form of a logic structure in order to of being provable. Then, the methods of confirmation of a theory should include combined logical, mathematical and empirical methods.

Let's start with the general outlook on our methods of study in physics. The observable terms, which are expressed through dimensional parameters, can be measured and obtained experimentally. On the other hand, the theoretical terms refer to assumptions, which connect terms with common definitions and universal laws in physics. The new theory of elasticity was built based on these premises. For example, if we apply an external force to a beam we can measure this force and deformation of the beam. The external force belongs to another body and therefore it is not a force in the equation of deformation. In the theory of elasticity the function of deformation contains the invisible elastic force that we cannot measure. It is necessary to build a chain of different physical assumptions and logical arguments in order to calculate a value of elastic force through other measurable or anticipated characteristics. Another example of a theoretical concept is the rate of change of a function (dY/dX). We can calculate the rate. This calculation includes an assumption: If proposition $Y=f(X)$ is correct, then the derivative is also correct. But, it is impossible to obtain the value of rate of change experimentally. We need to know the rate of a process in order to predict the behavior of an entity. The knowledge of the rate of change gives us an opportunity to make a comparative analysis of the structures and to see their future. Though our theoretical description of the relationship has variables which cannot be tested experimentally, it doesn't mean that a theory cannot be tested and be proved. A theory is a logical-mathematical construction and some of the criteria of correctness are logical rather than empirical. Combination of logical-mathematical methods of justification with empirical methods allows us to justify a theory or prove it wrong.

Another point we need to consider is that some of the experiments are very invasive. This especially concern experiments involving elemental particles and structures. In order to find the constituting parts of a nucleus, for example, we have to destroy it. The energy that is needed for such destruction is very high. It destroys not only the structure of a nucleus as a whole, but parts of it as well. Although one may find methods to see fragments of particles, we should not assume that those are the parts that exist in the formed matter under ordinary conditions. A process of destruction of matter finds different technological applications, but it could be the wrong method for understanding formed matter. Rather in theories, we deal with the holistic outlook on physics. We describe change of a whole in relation to the selected variable characteristics of the whole. On the contrary, the studies of physical structures are atomistic. We physically divide a whole in to its parts and study the characteristics and behavior of those parts. The idea, however, that the knowledge of the parts provides the knowledge of a whole is incorrect in essence.

It is possible to study characteristics of elemental particles in experiments not involving destruction. For example, experiments with light beams demonstrate the properties of a whole so and its parts. In spectral analysis and in quantum experiments physicists study properties of the parts of a beam of light. In these studies we have to accept the fact that not all properties we observe in the experiments of a whole belong to the geometrical whole.

Some of the observed properties may belong to parts of the whole. It is important to distinguish them. In some cases such analysis is simple and does not require complex suppositions. For example, suppose a body changes its habitual path in the space. The common supposition is that this body met another body or form of energy that caused a temporary or permanent change of its path.

Another example may be of force acting on a body that changes the geometry of the body. This change causes internal elastic forces to be distributed in the material of the body. We have an idea why the elastic force appears. We may equate the action of an elastic force to the action of an unbalanced external force, though we don't see the internal force and cannot measure it directly. We can do that because we accept the law of conservation of energy in physics. This law provides an opportunity to justify the observation that the removal of an external force on a body returns it to its initial position. The external force causes a hidden potential force to appear in the deformed body that is equal in value and opposite in direction to an external force. Elastic force acts when the equilibrium of an external and internal force is broken. Though we don't know the material essence of potential energy, we do know that it is energy which is associated with the displacement of equilibrium and distortion of space. We can calculate elastic force with the equation of equivalence of the external and internal forces. A formed whole has the ability to preserve its identity to some limit while changing its form. If we would be able to find this limit theoretically without demolishing every structure it would simplify and improve our knowledge. "The non-linear theory of elasticity and optimal design" contains the description how to calculate an individual limit.

Summary: Physics studies the properties of wholes using experimental methods. But, physics has no direct measurements for the forces of elasticity, inertial, or gravitation. Also, the empirical approach to the studies of properties does not provide objective methodology for constructing a theory. An experiment, even if it doesn't cause any doubt, is not a confirmation of a hypothesis. An experiment shows only concrete data pertained to concrete event under certain conditions. A general predictive theory is built by combining experimental and logical-mathematical methods. It is important to know the purposes and potential of experimental methods in physics. Experimental methods have limitations in establishing true relationship and predicting future events..

Mathematics in Physics

"The class "relations of ideas" comprises the *a priori* propositions of logic and pure mathematics, and these I allow to be necessary and certain only because they are analytic....They do not make any assertion about the empirical world, but simply record our determination to use symbols in a certain fashion. Proposition concerning empirical matters of fact, on the other hand, I hold to be hypotheses, which can be probable but never certain." (Language, Truth and Logic, A. J. Ayer)

Mathematics in physics is a scientific tool that should serve the purposes of physics. In its essence a physical equation can be distinguished from a pure mathematical equation. Mathematics uses abstract symbols. In a physical equation we deal with symbols which have physical meanings. Not only do symbols in a physical equation have meaningful concepts, but the connections between the concepts have a physical purpose, as well. These conditions put restrictions on the rules of mathematical operations in physics. Physical theory is usually presented in the form of mathematical equations. The appearance of physical ideas as unified with their mathematical forms overshadows the essence of real relations between mathematics and physics which should be investigated.

Let's consider the relations between mathematics and physics, keeping in mind that mathematics is a scientific tool in physics that serves physical purposes. Thus, mathematics allows for the substitution of symbols in an equation using the appropriate mathematical expressions. In on the second page of books on calculus one may find that a function of a function can be presented as a single function. In this book is the point of view that, principally, there is no adequate way to substitute in physics. A physical symbol in a physical equation is physical concept, which is involved in the unique conceptual relations with an entirely other concepts in a physical equation. The concepts and structure of a physical equation need to be carefully selected in order to have a meaningful propositional equation. Furthermore, in order to prove a hypothetical proposition logic requires testing not a proposition, rather the inference from a proposition. Then, one of the requirements to a structure of the propositional equation is that in order to have the meaningful inference the propositional equation should be non-linear.

One of the purposes of a physical theory in mechanics is to predict the possible geometrical change of the investigated whole, or change of its position while an unbalanced force is acting on the whole. For considering the changes we have laws which are our necessary general assumptions. All particular cases of change should comply with such laws. Thus, in classical mechanics acceleration of a body is proportional to the unbalanced force acting on the body and inversely proportional to the inertia of the body. The description of this law was made with the non-linear equation $a = F/km$, where 'a' is acceleration, 'F' is an unbalanced force and 'm' is the mass of a body that is measure of a body inertia. Acceleration has a non-linear relation with mass of a body. In the theory of elasticity change of geometry of the body 'Δ' is proportional to the elastic force 'F', and inversely proportional to the resistance of a body to deformation 'ER', $\Delta = F/ER$. Change has a non-linear relation with the resistance of a body to change, and therefore, this relation must be represented with a non-linear equation.

The assumption, which has existed in physics since Newton invented calculus, is that a whole can be presented as a sum of infinitesimal parts. This assumption is not congruent with the law of change, which shows relations between the physical characteristics of a whole. A non-linear relation cannot be substituted with a linear relation. Non-linear function has important properties that a linear function doesn't have. Thus, a non-linear function allows us to calculate a limit of the described relations. A linear function doesn't have a differentiable limit. The physical relations described with the equation of change have a limit, and therefore, should be represented adequately with non-linear functions. The complete non-linear definitive description of change has two equations, namely, the basic equation and its derivative. According to the rules of logic, for proving or disproving a hypothetical, basic, conceptual equation we do not have to prove the basic equation, rather the inference from this equation. The problem is that a derivative equation has a relative value as its result. The relative value can be calculated, but it cannot be measured. We cannot validate a physical theory using only the experimental method. For proving a theory it is necessary to compare the derivative function with a universally known function, such as tangent function. Also, it is necessary to show that the basic equation does not contradict the universal law of conservation of energy and momentum, and validate the basic equation experimentally.

The other point here is that it is impossible to predict the behavior of a body without comparing it to other bodies in similar circumstances, namely, acted on by the same unbalanced external force. Therefore, in making a prediction we do not consider change for an individual object, rather for a family of similar objects distinguished by their inertia or resistance to change. The curve of change representing such a family (see graph at the end of the book) needs three parameters to determine the position of an individual object in the family. It should be noted that through one point in a plane representing an individual object there can be drawn numerous curves. In physical reality only one concrete behavior of an object in the same circumstance is possible. In order to reach certainty in our predictions we need two coordinates of a plane and the tangent to a curve describing a possible behavior of an object.

The system of equations representing the law of change needs to be tested. According to logical demand we have to be certain that inference is consistent with the logical structure of the description. The mathematical deductive method gives us mathematical, rather than physical, certainty, by connecting the basic equation of a change with its derivative. Function for a family of objects, distinguished by characteristics of inertia, can be a curve for a particular force. Such a function has the derivative that forms a tangent function. The values of such a derivative are known for each particular point of a curve. The derivative equation is a logical-mathematical construction that has a relative value as its result. Thus, if the function of change is $\Delta=F/ER$, then the partial derivative is $d\Delta/dR=-F/ER^2$. This derivative can be calculated, but not measured. The result is correct if the basic equation is correct. The basic equation of change, on the other hand, has the absolute values of variable-concepts that can be measured. A basic equation can be tested experimentally. The problem is that we can measure values of the concepts, but we cannot measure the relations. The same result of a particular change can be obtained with different, so called, equivalent descriptions. The point of view in this work is that there are no equivalent descriptions in physics. Only one description is correct. The only part of a general curve of change that can be tested experimentally is the part of the rapid increase in the rate of change, if it corresponds to the rapid physical changes of an object.

Mathematics can describe any physical hypothesis. However, that doesn't mean that a physical hypothesis is necessarily correct, or that the mathematical means are selected properly. A theory needs to be tested. However, an experimental test of a theory is not enough to justify it. Numerous, repetitive tests can provide the same result for a conceptually misleading theory. For the complete testing of a theory it is necessary to have a logical-mathematical consistency of the theory. In order to have confidence in a theory we need to analyze the relations between a mathematical form and its physical content. A selection of the mathematical equations in physics depends on the substance of physical concepts and on the purpose of a physical theory. Mathematics can be viewed as one of the instruments for obtaining physical purposes. For the selection of mathematical equations we have to test the logical structure of a theory and the relatedness of a mathematical description to the physical purposes of a theory.

Mathematics by itself cannot guarantee the correctness of a physical hypothesis. Mathematics has no criteria for evaluating physical ideas. We need logic, along with its methods of justification, for proving or disproving a physical hypothesis. The idea that logic is a part of mathematics is confusing. Logic has methods and purposes that distinguish it from mathematics. Logic has two methods of justification, i.e. the inductive and deductive methods. Logic of pure mathematics is different from logic in physics. According to Bertrand Russell, in mathematics inductive logic is a proceeding which leads from previous events to the next. Bertrand Russell's definition of mathematical induction is as follows: "The principle of mathematical induction might be stated popularly in such some form as 'what can be inferred from next to next can be inferred from first to last'." According to the methodology in this work, building a physical theory is not based on establishing a connection to the first cause, rather on the connection of change to the related properties of a whole.

In physics the inductive method, which has justification, is a proceeding of a general law to the particular implication of this law. If the law of conservation of energy and momentum is accepted for all closed systems, then it is applicable to each particular closed system. Methodology in physics and methodology in mathematics are different. The most general obvious distinction is that mathematics is abstract and physics is concrete. There should not be an assumed direct connection between a physical hypothesis and mathematical forms and procedures. We have to justify our selection of mathematical equations and procedures. The role of a mediator between mathematics and physics can be fulfilled by logic and its methods of justification. Without logical structure it is impossible to build and prove a physical theory.

Logic has two main methods of justification, i.e. inductive and deductive. In classical logic the inductive method, as a rule, refers to a generalization made from repeated facts. In the new non-linear theory of elasticity, and in this work, the reliable inductive method refers to inferences from a general law to the particular statements which are under the umbrella of the general law. In physics if a law is, in general, true then it is true in all particular cases. Thus, all mathematical equations describing the relations among the physical concepts should comply with universal laws, such as the laws of energy conservation and conservation of momentum in closed systems. With the inductive method we test a selection of terms, and the consistency of the mathematical structure of a theory, on their conceptual correctness.

The deductive method in physics is identified with the derivative mathematical procedure from the basic conceptual description to the derivative equation, which is a mathematical conclusion. The derivative equation describes the rate of change of a non-linear function due to a change of its independent variable in the basic propositional function. The incompleteness of the mathematical deductive method in physics is that we have no method for proving the correctness of a propositional equation while the logical argument is that the conclusion is correct if the proposition is correct. A deductive mathematical procedure by itself cannot compensate this methodological insufficiency. Logic may help to construct a provable logical–mathematical system of equations, i.e. a basic equation and its derivative. Such a system is complete, and by testing both parts in combination with the inductive method, deductive method and experimental method, we may arrive to the conclusion on the correctness of a theory or prove it to be false.

Mathematics operates with symbols. Physics also accepts this convenient concise way of presenting physical relations among selected physical concepts. But, the symbols in physics are fundamentally different from the symbols in pure mathematics. In mathematics symbols are abstract. In physics symbols carry on physical content. The terms in a physical equation should have a concrete and exclusive physical meaning corresponding to the conceptual connections among them. Pure mathematics, on the other hand, is not concerned with the meaning of the symbols; rather, it is interested in the rules of operation of the symbols. However, if we take the rules of mathematics with no consideration of the differences between mathematics and physics, it may cause and cause misconceptions in physics. The necessity of a critical analysis of operational rules of pure mathematics has been demonstrated in the example of the new non-linear theory of elasticity.

A basic mathematical equation gives a hypothetical conceptual description of the physical relations in a process of change, namely, change is proportional to force causing or generated by change, and inverse to the resistance of a body to change, or to the inertia of a body. It is one of the basic laws in classical mechanics. A mathematical description has to follow this law. The mathematical function of change in the general theory of elasticity that refers to this description is $\Delta = F/ER$. The terms for this basic equation were selected by having in mind the uniqueness, and at the same time universality, of a physical event that is under the umbrella of this general law. The rules of mathematics allow us to substitute the physical terms in an equation with a seemingly equivalent description. The rule of calculus is that the "function of a function can be presented as a single function." Although mathematics allows for such substitution, by substituting the terms in a physical equation with other terms, we change the relations among the terms and change the function. For example, substituting term geometrical stiffness for R=kA/L we change an assumption that change (Δ) is inverse to a resistance (R). For Δ=FL/EA. change is now proportional to the length (L), and inversely proportional to an area of the cross-section (A). Although in simple cases, the resulting change Δ might be the same for both of these equations, the relations between the variables are different. The basic non-linear equation is still an incomplete description of the physical non-linear relations. However, the aforementioned substitution, without referring to a geometrical resistance as to a whole concept, results in a different conclusion about the rate of change in a process in the family of structures distinguished by the area of cross-section 'A.' Differential inference is a part of the description projecting the future. Without a proper identification of the independent variable it makes it difficult, or impossible, to arrive at a proper conclusion. Thus, the differential equation with R as an independent variable is $d\Delta/dR = -F/ER^2$ or $d\Delta = -FL^2/EA^2$. In the case with 'A' as the independent variable after substitution the rate of change of deformation is different, $d\Delta/dA = -FL/EA^2$. The proper selection of physical variable-concepts in a basic equation, without changing them throughout the logical-mathematical operations, is important for arriving to a proper conclusion. Mathematical laws of operation may help us to develop a physical theory under the condition of logical and physical supervision.

The same reasoning should prevent a combination of the parts of conceptually different terms in physical equation. It is often the case that a part that belongs to a force or moment combined to the geometrical concept of resistance to change. For example, in the classical theory of elasticity the vertical displacement of a beam commonly written, $Y_{max}=PL^3/48EI_x$, combines parts of concepts of different origin, namely, $L^3/48I_x$ has the part of elastic moment, $M=PL/8$, the part of the resistance due to geometry, $R=I_x/L$, and the part of resulting deformation, $\theta=6Y_{max}/L$. This equation of displacement distorts relations between meaningful physical concepts. Mathematically, however, it works for some purposes, and the displacements of a concrete beam along its length can be calculated.

Abstract mathematical structures and proceedings have connections with geometrical images. These images may suggest some physical ideas and uses. Geometrical forms of functions may help to select meaningful equations and proceedings in physics. Thus, linear function is represented with a straight line. The property of a straight line is that it has no derivative distinctive from this line. A straight line also has no differentiable limit. Every smooth curve in a small interval can be identified approximately with a straight line. However, complete adequate description of the physical relations, is non linear equation. The idea that a curve representing a physical function at any small interval can be presented with a linear function, and that the total change of the curve can then be described with the sum of those infinitesimal parts is in many applications a misleading idea. This idea leads to the misrepresentation of non-linear physical relations. The analysis of the properties of a function allows us to select a description with a system of non-linear equations, namely, the hypothetical conceptual basic equation and the partial derivative from that basic equation. A geometrically non-linear physical equation is a curve. Each point of this curve in a plain is characterized by absolute values of coordinates and its tangent, which can be obtained with a differential equation.

A whole in physics, in general, is not a sum of its infinitesimal parts. There is no functional relation between a whole and it parts. Functional relations can be established among characteristics representing conceptual aspects of the change of a whole. In a whole, change of one of its characteristics is intimately connected to the change of another of its characteristics. The relations we choose to describe are nonlinear because that is the essence of our understanding of a physical change, i.e. change of whole has limit. A change is proportional to the force causing it, and inverse to the innate property of a whole to resist a change. Or, the change of a position of a body is proportional to the force, and inverse to the inertial mass of the body. These relations can be revealed in the process of change of a whole, or in its position.

Nonlinear relations have a limit. In order to establish a limit we have to know rate of change at a given point, or in the interval where rapid increase of change occurs. Rate of change is identified with a tangent taken at subsequent points of a curve. An indicator of limiting change can be anticipated because we know the property of the tangent function, including the property of the tangent function at the interval of rapid changes. Prior mathematical knowledge of the character of a tangent function, which is associated with the rate of change of a function, helps to establish a limiting rate. A non-linear description of a process allows us to establish a limit for a process described with the function of change.

While selecting an independent variable, we shall keep in mind the physical nature of a theory. An independent variable plays a special role in the process of analysis of change. In the derivative equation the rate of change is a ratio of change over an independent variable. This ratio can only represent a physical event accurately if an independent variable that has the same units as its function is selected. This means that units should be comparable not only mathematically, but physically, as well. Otherwise, the relation between a function and its independent variable would be distorted. Just as it is improper to measure the length of an ant using a yard stick, it is also inappropriate to measure the change of the length of a beam that is in mm with the length of this beam. However, such practice often takes place in math, presenting physical relations which, indeed, lead to the distortion of relations. Physics, logic and mathematics dictate what kind of independent variable should be selected to be well-matched with a function. However, no attention is paid in physics to the dimensional compatibility of concepts.

Most physical functions can be adequately presented with non-linear functions of change rather than with a non-differentiable linear function, as it now stands. On the other hand, the movement of an individual body can be presented with a linear function. One of the properties of a line is the lack of mathematical limit characterizing a physical process. A non-linear function describes a curve. In order to have a definitive solution, such a function should be presented with two equations at different logical-mathematical levels. One is the basic equation which gives absolute values for every point in a curve, the other, a derivative equation that describes the relative change of a slope from one point to another. Such relative values correspond to the value of a tangent drawn at each point of a curve. Note, that the derivative function for a non-linear equation describing changes of a single entity is not necessarily the tangent function, while rate of change in the equation of change, which belongs to a family of entities, is the tangent function. In the theory of change the derivative function is an independently known universal tangent function. In this case, if a corresponding basic equation is supported experimentally in all instances, then we may conclude that such a system of equations is justifiable. Existential truth is unimportant in mathematical constructions, but it is important in physics. A system of the equations yields predictable results if the basic equation, supported by facts and its derivative, forms a tangent function.

Mathematical descriptions in physics logically follow physical purposes. Math in physics is not that independent from its rules of operations. Thus, the selection of a mathematical description depends on the physical purpose of a description. For example, if a purpose of a description is to find the length of a winding shoreline of a river, rather than describe some process, we may divide this line into parts and then find a sum for the total length. Another example of using addition is the method in quantum electrodynamics that summarizes the vectors. The vectors not only show possible direction of movement of a photon or an electron in a medium, but also the probability of such movement. A vector is a mathematical-logical construction. A result can be tested at the borders of the mediums for the incoming and outgoing photons. No theory truly supports such operations. Currently, physics mainly operates on the principle of equations summarizing changes of infinitesimal elements. Mathematically those descriptions are linear. A linear description has no meaningful inferences. A linear description may describe the present state for a singular entity, for example, linear movement of light. A linear description cannot project the generalization of behavior for different entities. Theory should be generalized by common law rather than by specialized method.

The observation and description of a singular event doesn't yield general predictions. A theory, on the other hand, has a purpose to predict future behavior for different entities. The description of a change of an object, or change of its habitual movement, requires a non-linear description which refers to a class of objects. The most general description is that change of an object is proportional to force acting on the object, and inversely proportional to the resistance of an object to change. A physical event can be the change of direction in comparison with non-obstruct theoretical movement. For ordinary objects, the more resistance/inertia an object has, the less change is expected. A photon has no mass, and therefore shouldn't have change in its movement. But if change occurs, then, hypothetically, this means that the photon attached itself to some particle that has mass and has inertial force that changes its direction. The mathematical description for such a model of events has to be non-linear. The derivative equation in a nonlinear description predicts behavior of an entity among similar entities of different mass. Non-linear description is also good for testing hypothetical models of physical events.

When visualizing properties of a function, one may see a corresponding curve in a plane. (See illustration at the end) Through a singular point referring to an entity, one may draw an unlimited number of curves in the same plain. Each curve has a different tangent at this point, describing a different rate of change, which results in a different behavior of the entity. In reality, the behavior of an object does not depend on our theories. Only one of the possible descriptions might be correct. For predictive analysis we have to select an independent variable that is dimensionally compatible, meaning it has the same increment with the result of the function. Variables in a function describe the most general conceptual characteristics of a whole and event. These concepts should be consistent in the course of analysis. The terms of a physical equation should relate to the object in which an effect takes place. The most general equation of change describes a universal physical law in the most general terms. Change of an object is proportional to a force causing the change, and inverse to the resistance of the object to change, due to its geometry or mass. A properly selected term of resistance has the same order of magnitude for its values as does a change.

Mathematics has no means to justify our hypotheses. It is the role of logic, through its inductive and deductive methods, to justify or reject a hypothesis. The deductive procedure belongs to mathematics. It is at this point that mathematics coincides with the means of logic. The main logical argument states that if a propositional equation is correct, then the conclusion is also correct. If we hold the position that reliable means of proof in physics are only empirical, then we have no way to prove a physical theory. We principally cannot prove the experimental correctness of a derivative mathematical conclusion, for it has a relative expression as its result. A relative value can be obtained mathematically. The only part of a curve that may have experimental support is the part where a transition occurs from the smooth linear branch of curve with a small incline to the almost vertical branch with the rapid increase in changes. If the transitional part corresponds to physical rapid changes, then the limit can be found with the tangent in the transitional part of the curve, and then the values of the tangent provide mathematical proof for all derivative values. A corresponding basic function, which is described with terms that have absolute values, can be tested and supported experimentally. In such a case the system of these connected functions are justified. Mathematics in physics undergoes some changes to be congruent with the demands of physics.

Physical theory is usually written in the form of mathematical functions, and uses pure mathematics rules. The current use of mathematics leads to conceptual flaws in physics. For instance, physical relations are usually presented with linear differential equations as the sums of infinitesimal parts. Actually non-linear relation in a function of change was presented as linear only because the independent variable was selected without taking in consideration the physical meaning of it relation to change. In pure abstract mathematics there is the rule that a function of a function can be presented as a single function. A non-discriminative use of this rule may and did lead to conceptually wrong theories in physics. We have to select an independent variable in a way that allows for non-linear relations to be presented as non-linear. For that, the independent variable should relate physically and dimensionally to the changes described in the function of change.

Summary: Mathematics in physics should match purposes of physics. The purpose of physical theory is to predict possible changes of whole by connecting change with the intrinsic property of a whole, namely, its resistance to change. The relation between change and resistance to change is non-linear and therefore it has limit.

Limit can be found with a non-linear function. Linear function has no differentiable limit. The behavior of object depends on its position in a curve describing force - change – resistance relationship. Relative position of whole in a curve is described with tangent to curve at this point. System of non-linear equations, basic and its derivative can be proven. The method of justification of derivative equation is inductive, namely equation should correspond to universal function. Basic equation, on the other hand, has to be supported experimentally. A meaningful logical-mathematical system allows us to build physical theory that is provable by design.

The Basic Concepts in Physics

"On the other hand, the scientific concepts are idealizations; they are derived from experience obtained by refined experimental tools, and are precisely defined through axioms and definitions. Only through these precise definitions is it possible to connect the concepts with the mathematical scheme and to derive mathematically the infinite variety of possible phenomena in this field." (W. Heisenberg, Physics and Philosophy)

For unifying methodology in physics it is necessary to have an agreement on the meanings of main concepts in physics, such as matter, space, time and force. Historically, the meanings of those concepts in science undergo changes. Thus, in Aristotelian philosophy matter was considered in its relation with form of existence. In the Greek period and until the 19th century the term "matter" simply meant material from which things are made. In Quantum Theory matter is considered to be formed by some previous chain of events, and the relation between form and matter is not considered to be of a particular interest.

"The "matter" of this period is "formed matter"; the process of formation being interpreted as a causal chain of mechanical interactions; it has lost its connection with the vegetative soul of Aristotelian philosophy, and therefore the dualism between matter and form is no longer relevant." (Physics and Philosophy, by Werner Heisenberg)

In this work, and in my previous work, *Non-linear theory of elasticity and optimal design,* the relation between matter and the geometry of a whole is considered to be very important for understanding the behavior of a whole. Change of the geometry of a whole causes tension in the material of the whole. The relationship between change of the geometry of a whole, resistance to change and induced tension can be presented as a function. In such a way, the properties of a whole, in regard to its geometry, can be evaluated.

Concepts in a physical equation can be divided into those that are obtained experimentally and those that are theoretical. Concepts that can be measured are empirical concepts. Concepts, that cannot be measured directly, and need connection to universal laws for establishing their values, are theoretical concepts. Let us consider some common concepts in physics. All physical concepts describe and belong to wholes. Physical theories are constructed in a frame of the universal law of conservation of mass-energy, which refers to closed systems. Every whole has its own space. For example, it can be space occupied by an atom, an atomic molecular structure, a macro structure of material, a common size body, or a celestial structure. The units of measurement of these spaces are different, but physical laws, including the law of change, are general.

Matter can be defined by the mass of an entity. In physics we are not interested in the chemical composition of matter. Rather, we are interested in the mass of a whole. Mass is sometimes identified with weight. In physics, however, mass is the measure of inertia of a whole, or it is the measure of resistance of a whole to change of its established velocity. Mass by itself does not give an explanation for the properties of a whole. The same mass may occupy different space, or have a different structure and the different properties. Matter and space it occupies in the process of comparative analysis allow determine resisting properties, strength and longevity of a whole. In the theory of elasticity the purpose of the theory of change is to establish a relation between the change of the geometry of a structure, the geometrical stiffness of a structure, and the stress distributed in the material of a structure. In classical mechanics of the rigid bodies change of velocity of wholes depends on the force applied and mass of a whole. In Einstein's Special Theory of Relativity the mass of a whole increases as velocity increases. Following the mathematical reasoning of the theory of change there is a non-linear relation between the mass of entity and increase of its velocity, since inertial mass represents resistance to change. A non-linear relation has a limit. If it is a general law then it should be true for all objects. In the case of particles with infinitely small masses and great velocities, the limit of velocity is the velocity of light in the vacuum. However, for entities of a common size and mass the limiting velocities are different. They can be calculated from a system of equations representing the law of change.

Space in Newtonian mechanics is "a passive container for matter and energy is absolute space."(I. Asimov) In Einstein's version space is curved with geometrical patterns in it. Einstein explained gravity by the geometry of space. In this work author has geometrical explanation for distribution energy in space that in turn explains gravity. Energy correlates with the local geometry of space. Space has the attributes of the physical existence. Space occupied by an entity is described with geometrical parameters that can be measured with linear and angular units. Geometry of an entity affects its properties. Dimensions of an entity have units of measurement, which are different from units of parts. Space of whole is occupied with parts situated at the different dimensional and energy levels. In the Special Theory of Relativity space measurements have values that are relative to the observer, or depend on the observer's movement. The Theory of Relativity allows make corrections of observations in different systems.

Time is one of the concepts we use to describe the duration of a process. In classical physics units of time are considered to be absolute values. In order to set accurate units of time physicists use periods of different simple harmonic vibrations and motions. These can be the period of vibration of an atom, the period of the earth rotation around its axis, or the velocity of light in a vacuum. Periodic processes and quantum behavior of energy are physical foundations of time measurements. Time units in different physical systems are different. The physical basis of time measurement changes in systems moving with different velocities. In Einstein' Theory of Relativity time is one of the space characteristics. In many theoretical descriptions of physical events the concept of "time" doesn't appear. For instance, the description of static deformations due to change of a body's stiffness doesn't include time.

Force is an important concept in physics. While energy is associated with the space geometry, force is associated with the interaction of entities and changes. If there is no change then no unbalanced force can be detected. Force is a theoretical concept. It can be determined through other measurable concepts or by calculation using the universal law of conservation of energy and definition of work. Theoretical nature of concept of force doesn't mean that concept is metaphysical. Force corresponds to empirical concept of geometrical change of an object or its position. Force is a part of a definition of work. By definition work is a product of force and change.

Laws in Physics

"[W]e shall see that the "laws of nature", if they are not mere definitions, are simply hypotheses which may be confuted by experience." (Alfred Jules Ayer, Language, Truth and Logic)

Laws in physics are our conditional boundaries of physical theories. Some general laws make it possible to create a common methodology in physics. All physical laws are assumptions that have no proof, but do have empirical support. There are two laws in physics that are considered universal basic laws. They are the law of conservation of mass-energy and the law of conservation of momentum in closed systems. All other specific laws in physics should be in compliance with these universal laws. Although we have no proof that matter-energy of the universe is constant and does not increase or decrease in time, rather we have proof of decay of energy in a whole, but in order to consider physical problems we have to start with an assumption that energy of a closed system is constant. Potential energy of a whole can be increased or decreased in the process of contact with another mass/energy. After the universal laws are established all physical models and ideas about how the universe works, and how its space-matter-energy is organized, should be in compliance with the universal laws of conservation of mass-energy and conservation of momentum. Both laws are the basis for unification of knowledge in physics. All physical theories, specific laws and definitions of the physical concepts should be in agreement with these universal laws. From this point of view let's consider the main laws of classical mechanics.

Newton's First Law of Motion is the proposition that a body remains at rest or in linear uniform motion with a constant velocity unless an external unbalanced force acts on the body and changes its velocity, namely changes its speed and/or direction of movement. Analysis of this law shows two propositions. First, is that uniform motion in a straight line and with constant velocity does not require additional force for a body to continue such movement (neglecting friction). This idea is in compliance with the law of conservation of energy and momentum. The other idea is that the change of the linear uniform movement of a body would require contact with an outside unbalanced force. The law, also, suggests that rotational motion of a body needs additional energy from an outside source for rotational motion is undergo changes. In the universe we observe bodies that have permanent rotational movements in orbital curved paths.

The question arises: Do the laws of conservation of energy and momentum concern only bodies in linear movement, or are they general laws that can be applied to bodies with different patterns of movement, as well? The conclusion here is that there is no additional force required for a system that reaches equilibrium of forces. It would be necessary to apply force to a body in order to change the body's orbital path. The laws of conservation of energy and momentum are applicable to all bodies that have established patterns of movement in any system of bodies. There is no empirical or logical explanation why bodies in rotational movement need gravitational force for continuing movement in their paths.

In the case of equilibrium of forces acting on a body, the body obeys the laws of conservation of energy and momentum. The history of created the systems of bodies in the universe may provide an explanation that is consistent with universal laws, but these laws were not consistently considered in modeling the universe. One of the explanations was offered by Einstein: The bodies are laying their paths and distorting the space. However, the elasticity of space demands the spending of energy for making a path. Another explanation is proposed by particle physicists. They attribute the unifying role of a universal force to some elemental particles. In this hypothesis classical mechanics has an unexplained force of attraction that keeps bodies at certain distances. Particle physicists attribute the attraction among celestial objects to the elemental particle named graviton. The graviton has not been discovered. There is another model in physics which connects bodies by the strings in a rope-like fashion in multidimensional space and requires energy for manipulation of bodies. Any idea that proposes spending energy for supporting a system, which already exists in equilibrium, contradicts the laws of conservation of energy and momentum.

The process of the creation of the universe is the process of creation of matter and space. Let's imagine a space that is already in existence with no force of gravity acting in the distance. We can make a hypothesis that changes occur in the historical process of the   creation of a system. Big masses exerted energy that is distributed in the surrounding space, probably, in a wavelike manner as it was suggested by Einstein. Those concentric waves of energy don't propagate or significantly fade in time in a system in equilibrium. The wave structure of space belongs to the system as a whole. The formed bodies such as planets are situated in the troughs of concentric waves of energy formation. In order to get out of its path a body should spend energy. There is a law of nature known as the principle of minimum total potential energy. This principle was formulated as follows: From all the known possible systems of movement of kinematics, only real movements give minimum potential energy. This principle prevents the independent movement of a body in the direction that increases its potential energy. The opposite is also true. A body needs energy from outside to climb from the bottom to the top of an energy wave in order to release itself from the predicament of its movement.

Our models of order in the universe cannot yet be proven. The only support we may have for a model it is to be in compliance with the universal laws, and be true to the facts of those laws. In the absence of unbalanced forces, after order is established, the order may remain indefinitely for the lifetime of a system. Only subtle degenerating forces acting in time can eventually bring another, order or disorder, to an isolated system or its parts. Analysis of forces acting on a body, spinning around its axis with some slope, shows a vertical in the direction of increase for a potential energy lifting force. This force is not big enough for lifting a massive body to the top of an energy level. Horizontal force, which is tangential to the curve of a path, moves a body in its path. This body is in equilibrium with its surrounding energy field, without the need of additional energy, and thus it obeys the laws of conservation of energy and momentum.

Force is an important concept in physics. Concept of force is associated with the change of a whole, or system. If there is change, then there should be an unbalanced force. If there is no change, then there is no unbalanced force in a system. Force appears to be a representative of action of one body on another for bodies in contact. From everyday experiences we know that we need to apply force only while assembling a whole out of its parts, or disassembling a whole into its various parts. Otherwise, no force is needed to keep a whole as it is. This refers to linear and nonlinear established movement of a whole, as well. In a common model, while demonstrating the movement of planets around the Sun, we attach an object to a rope and rotate it. As soon as we let the object goes free it travels for a while in a straight line. Such a demonstration of what might happen to a planet without constant supervision of the Sun is not a model that complies with the laws of conservation of energy and momentum. In this model energy of the Sun constantly should be spent on the movement of planets. Although, we cannot yet prove our models of universal order, at least these models should be consistent with our own laws for establishing a uniformity of models. Unless the law of conservation of energy is rejected, the physical models should comply with this law.

Many different models, which connect parts of systems, exist. For the selection of a model, however, we should keep in mind that if there is no change, then no unbalanced force exists in the system. We have no means of detecting forces that don't express themselves. Rotational or linear established movements are the natural order. Moving bodies can be considered independent entities that are in balance with the energy in their space. On the other hand, change of an established rotational orbital movement on linear movement would require force. Force is present in the process of creating or destroying a whole.

Gravitational force is a force which action can be explained with the principle of "minimum potential energy". There are differences in potential energy at different levels of space-structures of systems. According to the principle of minimum potential energy free natural movement of a body is always directed toward decrease of its potential energy. Not only large masses situated in a ditch of low energy in the space, but an apple falling from a tree also moving in the direction of low energy that is occupied by earth. Gravitational force is not a rope with different strengths for each object connecting it with other manipulating objects.

Inertia is another important concept in physics. A whole resists changes, whether it is the change of the geometry of a whole or the change of its path. This resistance of a whole to change is called inertia. In classical mechanics the mass of a body is a measure of a body's inertia. In mechanics of elastic bodies resistance of a body to change depends on its geometry and elasticity of material.

Newton's Second Law of Motion connects change of velocity of a body in the straightforward movement with the force acting on the body. A change of velocity or acceleration of a body is proportional to the force acting on the body and reversely proportional to the mass of a body. Newton's Second Law then represented with a general equation, $a=f/m$. Another possible change of established order is the change of the geometry of a whole and the change of place and path of a whole in a bigger whole. Any change of the geometry of a body or change of its place requires a force to overcome inertia. Classical mechanics does not consider established rotational motion of a body obeying the same laws as for a body in straight line movement. But bodies have inertia for sustaining rotational motions as well.

Newton's Third Law of Motion states that bodies coming in contact with each other exert equal in magnitude and opposite in direction forces. The third law is completely in agreement with our understanding of interactions in the physical universe. There is no action at a distance; the amount of energy exchanged in contact is the same, affecting each body. Although the bodies in contact are acted on by forces of the same magnitude, a change of each body is different. In general, the bodies are different and have different inertial mass. The third law has important logical and physical implications for analysis the physical events. In such analysis we separate cause from effect, for there is no functional connection between independent bodies representing cause and effect. We attribute the role of cause to one body in contact, and another body is the effect. Changes that occur in a body-effect can be described with a propositional mathematical function and its derivative. The function of change in mechanics is $\Delta = F/kR$, where 'Δ' is a symbol of change, 'F' is a force/stress that is generated by contact of bodies and 'R' is a resistance of a body to the particular change. The function of change is important for the definitive description of an event and predicting the behavior of entity. The same method of functional analysis can be applied to another body in contact, or to another event or phenomenon.

Mechanics of Elastic Bodies

The main physical idea of elasticity was discovered by the English scientist Robert Hooke in 1660, and published in 1678. All bodies/wholes, and all materials, possess elastic properties. This physical idea is general because term 'whole' can be referred to any formed entity whether it is a solid body, atomic structure, molecular structure or galactic structure. The new non-linear theory of elasticity has been developed on the material of structural engineering mechanics. The theory of elasticity is the theory of change of elastic geometrical wholes. In the mechanics of elastic body's geometry of a body changes while affected by an unbalanced external force. A change or deformation can be reversible/elastic if after removing an unbalanced force a body returns to its previous shape and position. Deformation can be, also, partially inelastic. In all cases, energy is spent on the change of the geometry of a body. Change of geometry occurs when a body cannot move freely in the direction of force. Most engineering structures are subjected to deformations. Engineering elastic analysis is necessary in order to design safe and reliable structures. Elastic analysis uses the basic equation of equivalence of stress per unit area (σ), and strain that is the elongation per unit of length (ε), for the infinitely small part of a structure. The equation, $\sigma = E\varepsilon$ (1), is named Hooke's law. However, behavior of a whole structure cannot be inferred from this equation of equivalence of stress and strain. It is not only matter that has elastic properties, rather space, as a container of matter, has elastic properties. Properties of structures depend on an organization of matter in space. A whole is not a sum of its parts. The law for infinitesimal volume is not a law that can be mathematically extended onto a whole structure. Classical mechanics misunderstood and misrepresented the relationship between matter and space. The traditional equation of the total deformation of a bar, for example, is obtained by substituting the concepts of stress and strain, in the equation of equivalence (1), with expressions that define stress and strain. Stress is a force 'N' per unit of area of cross-section 'A,' $\sigma = N/A$ (2); and strain is the total elongation of a bar, 'e' per unit of length, $\varepsilon=e/L$ (3). Thus, the standard classical formula was created, $e=NL/EA$ (4). Coefficient of elasticity $E=\sigma/\varepsilon$ is experimentally obtained on a standard size of a specimen. This coefficient is different for the different materials. Mathematics allows substituting symbols in equation with their equivalent expressions. In physics such a practice may lead to physical misconceptions. Relations in physics are concrete and exclusive. They should be represented with carefully selected concepts that provide adequate proven relations. For this it is necessary to create a non-linear description of change for a whole body, produce its derivative and test this system of equations. A system of equations that is based on the traditional equation of deformation has no experimental or logical support. One may find a detailed analysis in my book, *Non-linear Theory of Elasticity and Optimal Design.*

According to the new theory of elasticity deformation of structure is proportional to force distributed in a structure and inversely proportional to resistance of a structure to deformation. Resistance of a structure depends on elastic properties of material of a structure and the resistance due to the geometry of a structure. Structure has an individual limit of elasticity. It is a limit one needs to know in order to make safe and reliable structures. The problem is to find an individual limit with a non-destructive method. An elastic limit that demonstrates itself in the test of a structure can be of different origins. It can be a limit of elasticity of material which is obtained experimentally on a selected standard size of a specimen. This limit of material has statistical value. A limit may depend on the geometrical property of a structure that is called geometrical stiffness 'R.' This specific individual limit is calculated with the new method of optimization in the new non-linear theory of elasticity. The real limit of elasticity of a structure, according to the new theory, is the relatively lesser of those two limits: the limit of material and the limit imposed by the geometry of a structure.

An important part of the new theory of elasticity is that the different experimental conditions may reveal properties of a structure as geometrical wholes, or properties of the constituting parts, namely of material. We have to distinguish them. Both a whole and its parts may demonstrate their characteristics in appropriate experimental conditions. In experiments on whole structure we are unable to obtain the characteristics of the basic atomic-molecular structure. In order to find these properties we need different specimens and techniques. The purpose of such studies is also different. They are concerned with the basic structures of different materials. The methodology of electro-dynamics, hydro-dynamics and other kinds of dynamics, however, may closely follow the methodology of analyzed dynamics of elastic bodies. Engineering sciences are deficient without this new knowledge.

Light

In classical mechanics light is considered to be a phenomenon. However, light can be considered as an entity similar to other entities such as solids, fluids, and gases. Matter-energy exists in many forms. Physical properties of solids, liquids, gases and electro-magnetic radiations are different. But, the methodology of the study of the properties and behavior of a whole and parts of a whole is similar. Every formed entity has characteristics related to its geometry and to change. An entity belongs to a group/family with similar properties. Comparative analysis changes in a family are our tool of research.

A beam of light and beam made from steel, seemingly, do not have much in common aside from their forms. But, indeed, a beam of light can be considered as an entity moving in a medium. (Note, that now there are experiments which stop light.) A beam of light has an internal structure that is consistent and stable unless outside forces acting on the beam of light change these structures. A beam of light displays its properties as a geometrical whole. A whole consists of a hierarchy of parts. These parts have different properties from a whole beam. In some experiments we deal with properties of parts that demonstrate themselves instead of properties of a whole. White light has parts that have different frequencies. To the human eye different frequencies are seen as different colors.

The general properties of light are reflection, refraction, dispersion, diffraction and interference. These properties are obtained through different experiments. There are properties of light as a whole, and other are properties of constituting parts. Similarly to experiments with elastic behavior of objects, experiments with light do not show the behavior of whole and elemental particles constituting a whole simultaneously. For example, the strength of a structure made of steel in tensile experiments may show the strength of a geometrical whole or the strength of the macro-structure of material. Such experiments don't show strength of an atomic molecular structure, strength of an atom, or strength of a nucleus. The same is true for a beam of light. Our measurement of the speed of light refers to a beam of light as a whole. A beam of light, as a whole, travels with constant velocity straight and forward, unless its movement is affected by an outside force. All parts of a whole moves together. Properties of particles, electrons and photons, should be studied separately. The Theory of Relativity explains relativity of motion and independent characters of properties of parts. For instance, passengers in a train have the same velocity as a part of the train. At the same time, a passenger is independent in his movements in the train. The velocity of light cannot be explained by the velocity of its fundamental particles, such as electrons and photons. At the same time particles have independent movement-vibrations. General properties of light, depending on experiment, may belong to light as a whole or to parts of light.

Reflection of light beam from some surface is, indeed, reflection of light as a geometrical whole.

Refraction is the bending of constituting parts of light while light passes through a medium. Different materials have different optical density. Different parts of light, with different wave lengths, have different angles of refraction or a different Index of Refraction in different mediums. In such experiments, when light is divided into its different parts, we can study the behavior and properties of these parts independently. An individual part can be considered as a new whole.

Diffraction is the bending of light around an obstacle in its path. Bending depends on the size and geometry of an obstacle. The behavior of light in such experiments resembles the behavior of a stream of water having obstacles in its path.

Interference is the phenomenon of interaction of two or more energy formations. Such interaction obeys the rules of merging energy. The elemental particles in this hierarchy of parts of light have different frequencies. Parts of light from different sources merge if they have the same frequencies. Amplitude of such frequencies will be twice the original amplitude, and twice the energy. Even if it looks like the behavior of waves we do not deal with waves, but with particles that have different frequencies and occupy a certain place in the spectrum. Interference doesn't point to the double nature of light. It is rather a manifestation of the fact that parts have their own properties and under certain conditions may act independently. Interference means adding energy to a part of light as to an independent entity. Patterns of light structure remain the same; linear velocity of light as a whole also remains the same, but the amplitudes and energy of particles are changed. The behavior of a particle depends on its position in the spectrum of a light structure. As a whole, light beam has properties that are different from the properties of its elemental particles. These properties cannot be studied within the same experiment. We accept the idea that, in essence, the same elemental particles can be parts of different entities. We have to accept that the properties of a whole are different from the properties of its parts. A whole and its parts can be viewed as independent entities. No functional connection can be established between them. The change of a whole can be represented as a function of the selected characteristics of a whole. As a whole, light consists of parts that have different properties. Each of these parts of light also is made up of different parts. These parts have elemental particles that have different frequencies. Depending on their frequencies particles have a certain place in the structure of light.

The methodology of analysis of the phenomenon can be common for different phenomena. The properties and behavior of a light beam can be understood using the same approach as in the analysis of a beam of steel in the theory of elasticity. The study of a phenomenon usually refers to a family of entities. We cannot study experimentally a whole object and the parts of an object simultaneously. The properties that are obtained in such experiments belong to different entities, rather than indicate the mysterious double nature of light.

Quantum Mechanics

"All physical theories that did not take quanta into account, but assumed energy to be continuous are sometimes lumped together as classical physics, where as physical theory that does take quanta into effect is modern physics." (I. Asimov)

Quantum mechanics claims exclusiveness: exclusivity of its laws in comparison with classical mechanics; exclusivity of its philosophy; exclusivity of a theoretical presentation of events; exclusivity of its empirical methods; exclusivity of both its physical and mathematical space, which are different from our classical perceptions and views on nature. This uniqueness brings us to impossibility consolidating the principles of quantum and classical mechanics. Let's compare principles of quantum mechanics with the epistemological principle presented here.

Epistemological principles should be general, for they do not refer to a concrete phenomenon, rather to the methods empirical and analytical connecting theory to physical reality. One of the main principles of quantum mechanics is that energy exists in a form of quantum, namely, that energy comes and goes in discrete packages.

This view is in compliance with the atomistic view on the existence of matter. It also doesn't contradict the idea that energy can be considered a form of matter and vice versa.

Quantum mechanics studies elemental particles and interactions between elemental particles and quanta of energy. The interaction of quanta of energy with a particle can be considered a different kind of relationship, distinguishing itself from mechanical interactions. In a mechanical interaction an entity is not converted into another entity. Though some changes may occur, essentially the object remains as the same whole consisting of the same parts. In a quantum interaction quanta attach themselves to an electron, and may attach another particle and another photon. A new entity may appear as a result of such interactions. The description of such a process of creation of matter is different from the description of change of an entity. The separation of a purely mechanical interaction from a quanta-mechanical interaction may help to understand the phenomenon, and prove the mathematical-logical structure of a theory referring to quantum phenomena.

When starting a research, we separate cause from effect. Cause finds its realization in the effect. We separate cause-energy from an effect that is the change of an object of investigation. Physical theory does not describe the cause – effect relationship because there is no functional connection between them. We may see or may not see a cause of changes, but theory only describes an effect. Thus, a theory of change considers the following properties: change; energy participating in a process; and resistance to change, due to media properties and to the distance particles travel. In this work physical theory relates to a family of similar entities.

Quantum Electrodynamics

"The more you see how strangely Nature behaves; the harder is to make a model that explains how even   simplest phenomena actually work. So, theoretical physics has given up on that." ("The Strange Theory of Light and Matter" by Richard P. Feynman)

"Attempts to understand the motion of the electrons going around the nucleus by using mechanical laws – analogous to the way Newton used the laws of motion to figure out how the earth went around the sun – were a real failure…" (QED, Richard P. Feynman)

Quantum Electrodynamics deals with interactions of particles and, in particular, electrons. QED is not a complete theory of events in the micro world of elemental structures. But, QED has a particular method of study for the behavior of individual photons passing through substances such as glass plates of different thickness. This method allows calculate the probability of events by adding arrows of probability for every possible event. A suggestion for using this method contains reservation, as displayed in following citation:

"When I say that all the phenomena of the physical world can be explained by this theory, we don't really know that. But if we arrange in the laboratory an experiment involving just a few electrons in a simple circumstances, then we can calculate what might happen very accurately, and we can measure it very accurately too." (QED, by Richard P. Feynman)

Quantum mechanics is concerned with the properties and interactions of elemental particles of matter with quanta of energy. Elemental particles and quanta of energy are in the foundation of the hierarchy of structures in the universe. For obtaining knowledge of the properties of elemental particles and their structures it is often necessary to separate these entities from the more complex structures which they make up. Experiments on particles are very invasive. These experiments, practically, destroy matter. Some of these experiments are done at high speed and high energy levels. Matter in these experiments becomes more and more disintegrated. More and more elemental particles appear. In ordinary organized matter these particles do not exist. A condition of high energy of space is required to conduct these studies. But, there is no proof that particles, and forces associated with the processes of the creation of matter survive, at a low energy of organized matter when equilibrium of forces is attained. Quantum electrodynamics considers particles, and ignores the whole structure of light behind particles. "Light looked like waves at first, and its characteristics as a particle were discovered later."(QED, Richard P. Feynman) The result is a controversial point of view on the double nature of light. Light can be considered a whole object, for it behaves like a whole. Light has parts with some wavelike behavior. However, knowledge of the properties of elemental particles doesn't lead to the knowledge of properties of a whole. Every whole has its properties characteristic. Parts of a whole have their own properties.

Physics and Physicist

"So, in like manner, the intellect, by its native strength, makes for itself intellectual instruments, whereby it acquires strength for performing other intellectual operations and from these operations gets again fresh instruments, or the power of pushing its investigations further, and thus gradually proceeds till it reaches the summit of wisdom." (On the Improvement of Understanding, by B. Spinoza)

Physicist deals with problems in physics. A physicist studies physical structures of the universe and elemental particles, and creates hypothetical models of structures. A physicist also studies physical processes and creates physical theories.

Many definitions of the term "theory" exist. But, no matter what definition we choose, a physical theory is a creation of scientists. In the process of building a theory our philosophy and general outlook on structures of nature, forces in nature, time and space, and our specific knowledge, plays its role. We are self-critical enough to realize that our ideas necessarily have subjective elements. Although, in principle, we cannot get rid of the subjective elements of a theory, we can use some formal methods for the objectification of a scientific theory. The term 'objectification' means that we use the formal methods such as logic, mathematics and experiments for the designing, correcting and testing of an internally consistent theory. Although concrete theory is based on specific material, a common approach can be used when dealing with any material.

Physical theories are not copies of reality that describe with mathematical symbols hypothetical physical relations. It is necessary to have the rules for constructing and testing physical theories. Physical data is not only an objective component in theories. A theory is usually concerned with the description of change of an object, rather than with a physical object in its totality. A theory should have a logical structure and rules for its construction. The following prior knowledge should be considered: mathematics that is adapted to the purposes of physics; and data for a theory should be selected appropriately for its purposes - all that separates a theory from purely empirical methods of studies of the objects. A reliable theory cannot be built based on some physical data and our assumptions. Even if our ideas about structures and processes in the universe are correct, it is necessary to have the means to connect these ideas and knowledge with the means of analysis of data. For example, we can experimentally obtain data on change of the geometry of an object, in a particular case. This data does not yield a prediction of behavior of a whole under different circumstances.

In order to predict the behavior of an object it is necessary to see patterns of behavior that are described with a function for a family of objects. A basic function describes hypothetical relations among selected variables and, in particular, the relation between a change and an independent variable. Relative change that depends on change of an independent variable is described with a derivative function. In order to prove a hypothetical function it is necessary to prove an inference from this function as well. In order to prove a derivative function we may compare it with another universal function that has a similar pattern of relations. Such a method is used in my new theory of elasticity.

Nature doesn't lead us to conclusions about the future. We can only test specifics, and present events in the constantly changeable world. In a theory we utilize means that lead us to conclusion about a possible future. Methods of the testing of theories cannot only be empirical, for creation of a theory involves logic and mathematics, as well as selected facts. A theory describes relations among selected characteristics of an object. We have no means to test relations empirically. In order to have the possibility to test a theory we need a logical structure as the frame of theory. Within that logical frame we can test the correctness of selected concepts and the relations among these concepts.

When creating a theory we use subjective and objective formal methods. Thus, for conceiving physical ideas one needs general and specific knowledge, imagination, a critical mind for selecting ideas and facts, the ability to organize ideas and the ability to communicate and explain his ideas. All these are subjective means, depending heavily on the individual. We also need formal methods, such as logic, for selecting concepts and building a structure of a theory; empirical methods for gaining the facts and testing a theory; and mathematical methods for describing connections between concepts, making inferences and the mathematical transformations.

It is worth mentioning that logic and mathematics are not the equivalent branches of science as some scientists believe. Both subjects play their own roles, have their own philosophy, and maintain their individual philosophy and methods in the process of building a theory. We use methods of logic in the process of the objectification of a theory for selecting concepts, building logical structure of the theory, establishing limits of the theory and testing the structure of the theory. The term 'logic' is sometimes used in respect to subjective qualities of our thinking and rationalization process. However, such logic refers to the physiology and psychology. In that sense, logic is one of the qualities of an individual mind, and as such, it is subjective. We should not confuse it with logic that is a scientific method of justification. Objective logic has its own philosophy, concepts, rules and sphere of application. Formal methods of logic that test empirical and mathematical methods in theories are, indeed, objective.

Objective methods are also based on system of prior knowledge. While some system of knowledge is accepted, it is considered to be an objective knowledge. Laws, such as the law of conservation of mass-energy, are considered general objective knowledge. Changes in science occur when we deal with inconsistencies in theories, and a disconnect a theory and facts. The process of creating physical theory requires the governing of sufficient facts, and an ability to observe and analyze the facts. However, the observations and facts don't yet lead to a theory. The creation of a theory is a more sophisticated process. Our theories are not reflections of nature, or the descriptions of assumed connections in nature. A theory is, also, not a description of our understanding of nature. A theory has its own way to present nature, its own rules, and its own forms. The purpose of a theory in physics is to describe the relations of selected concepts/properties mathematically in order to predict behavior, and these properties, quantitatively. One of the purposes of this book is to establish the elements of formal steps for building reliable theories.

The principle difference between our *physical models*

of reality and *theoretical models* in theories is not only that in nature we deal with physical entities, it is also that in theory, we deal with symbols and models of relations among properties of entity. The division of a physical entity leads to the hierarchy of smaller and smaller physical entities. Analysis in a theory, on the other hand, leads to a description of relations among selected characteristics of an entity. For example, if we are interested in the property of strength of a structure existing in some geometrical form, the characteristic that we select is not the general geometrical descriptions of a body; rather it is characteristic describing resistance of a structure to deformation. The equation of change includes deformation, elastic force, geometrical stiffness and the coefficient reflecting elastic properties of material. These properties are not obvious through observation. They belong to our point of view on the process of deformation. If we study electrical properties of a body, we include the characteristic of resistance of body to the passage of electrical current, potential difference and electric force. All these physical concepts have a place in Ohm's equation. Specific properties of matter are counted with experimental statistical coefficients.

Physics studies structures and properties of matter at different levels of division, which include everything from the infinitesimal level to the whole itself. An entity or a whole consisting of matter and energy exists in some geometrical form. A whole has properties not only of matter it made up, but also the properties of space that this whole occupies. Specific properties of space are different from the properties of matter. Those properties depend on distribution of the matter and energy in space. The goals of physics do not only to understand structures and connections in nature, but also the creation of theories that predict behavior and limit of the existence of wholes. We have to prove the methodology that pertains to the building of theories in general, and while creating a theory to prove that particular theory, as well.

Our means for creating a theory are different from nature's means of creating an order to space-matter. The methods we use in theories belong to an organization of ideas and observations that stem from formal subjects such as logic and mathematics, and the process of experimentation. These methods cannot be found in nature. Not only are our ideas subjective, but our observations are subjective to some extent, as well. Different formal methods assist us in the creation of objective physical theories, which may give reliable predictions. Furthermore, our schematic descriptions of a phenomenon often correct our understanding of nature.

In modern physics the description of phenomena can be mathematical models that may not resemble relations in nature at all. But, these mathematical models may produce results close to the outcome in observation. For example, in quantum mechanics Heisenberg's quantum matrices give accurate results, although those matrices have no physical or schematic resemblances with the process to which they refer. Nevertheless, it is preferable to have physical explanations for theoretical models. In classical mechanics theories, such as the *Non-Linear Theory of Elasticity*, are constructed in a way that includes mathematical inference from basic conceptual description. Such inference has a pure mathematical value, which cannot be tested experimentally. Formal methods are used to prove such theories. Theoretical constructions and models in classical mechanics can be explained physically. Reliable results in a theory can be obtained with no reference to hypothetical structures of matter or hypothetical ideas about forces that keep matter together. Mechanical properties of a body, for all practical purposes, can be obtained without reference to atomic structures of matter. In the laws of elasticity, electrodynamics, aerodynamics and hydrodynamics - all are similar - change of the object under investigation is proportional to reactive force in an object, and inverse to the resistance of the object to change. This law of change allows the prediction of behavior of wholes under different conditions.

Man has a creative mind. We have the ability to select information out of the physical world, classify and store that information in the memory, connect related facts, conceive some concepts and ideas compare and make conclusions and remember and retrieve related information from the memory. The mind performs its operations on two levels, conscious and subconscious, and we are not always aware of how particular operations are carried out. The bases for a mind at work are the physiological processes in the brain that have educational background and patterns of knowledge. The work of an individual mind may produce creative, but subjective, ideas about the universe and probable causal connections in the universe. In order to build a scientific theory that is not subjective, we need to use formal methods that help us to convert hypothetical ideas into a well organized objective knowledge. The main differences between scientific theories and hypothetical ideas are that, first of all, theories have or should have consistent logical structure, and secondly, theories can be tested and proven or disproved. The process of converting a hypothesis into a theory can be called the process of objectification. For that purpose, in physics we use the methods of logic and mathematics, and the empirical methods of observation, experimentation, modeling and measuring. Although the sciences are also our inventions, and products of our minds, we make a distinction between raw materials, individual hypothetical ideas and their connections, and previously learned facts, and, on the other hand, organized subjects and their philosophies, methods, concepts and purposes. Combining these methods helps to build an objective foundation for reliable theories.

Summary: Scientific theories do not come to us via the processes of observation and contemplation. A theory is not a reflection of reality or a mathematical description of relationships in nature based on some observations. A scientific theory has a different purpose from an observation, and follows different rules. The purpose of a theory is to predict the behavior of an object in a process of change. We select characteristics of a whole that are related to change. And in turn, these characteristics tell us about a process, rather than about the structure of a whole. In the process of constructing a theory we use all means available: hypothetical ideas, a priori knowledge, logic, mathematics and empirical means. The processes of constructing and testing a theory have some rules that are not the laws of nature, but the laws of subjects that we use. However, scientific theory requires and yields laws that can be considered as laws of nature. In the historical process of acquiring knowledge some universal physical laws, such as the laws of conservation of mass-energy, conservation of momentum, equivalence of mass and energy and quantum mechanics, which explains that energy comes and goes in discrete quanta of energy, have been formed. Although it is not nature that handed us these laws, we call them laws of nature because we believe that these laws are necessary for our studies of nature, and in closed systems they are supported with facts. These laws become objective priori knowledge in the classical system of knowledge. All theories and derivative laws obey the main universal laws, and thus, physical theories are connected through a common system of knowledge. This doesn't mean that the system of knowledge has no room for critical analysis of laws and improvement of knowledge. That happens constantly in the history of science. The change of the methodology of science is more uncommon. However, the current crisis in physics demands a revision of its methodology. Communication between nature and a physicist consists of processing information through accessible means. Good theory improves the understanding of nature. New facts may support or destroy a theory. For the construction of a new theory, or the improvement of an existing theory, we need new methodology in physics. In a new methodology it is necessary to review the basic notions of physics, logic and mathematics.

Logic in Physical Theory

"There is no absolute standard of rationality, just as there is no method of constructing hypotheses which is guaranteed to be reliable" (A.J. Ayer, Language, Truth and Logic)

The term 'logic' refers commonly to an individual process of rationalization that is used by an individual in science and life. As A.J. Ayer said of such a logical process, "There is no absolute standard of rationality..." However, there is a method of constructing theories which is guaranteed to be reliable. Physical theories require objective scientific logic that has specific terms and definitions, specific logical structures and rules of operation. This work describes elements of objective logic, rather than the author's logical process of thought.

Theories in physics have some logical-mathematical structures. Only such a theory that has a consistent logical scheme has the possibility to be proved. The purposes of logic in physics are to justify a selection of concepts; justify a mathematical structure of a theory; and prove a system of an initially hypothetical, basic conceptual description and its derivative description. However, logic has not readily been available for its systemic applications. Let us consider some methods of logic for building a mathematical structure of a physical theory; there are rules for selecting variable concepts of a theory, and rules for justifying a theory that has consistent logical structure.

There is no logic in nature. Physical structures do not obey laws of logic, but rather, laws related to the distribution of mass/energy in space. We may create hypothetical models of relations among parts situated in the physical whole at different levels. But, these models cannot be proven, since no functional relation can be established between independent entities situated at different energy levels in space. Change at one level does not bring certain functional change to another level of organization of matter. However, if change of a whole occurs, then change can be described as having functional connections, which are related to the changing characteristics of a whole, rather than a connection with an assumed or even known cause.

Physical theory requires a logical scheme in its foundation that allows a theory to be justified, or to be proven wrong. A theory in physics is not a simple reflection of observations of a natural phenomenon, nor based on a conclusion from experiments. One or more observations of a particular event are not a reliable basis for making justifiable general predictions of the behavior for an object, or other similar objects. However, the physical behavior of a member from a physical-mathematical family can be predicted by using the process of comparison with neighboring members in their natural order. The purpose of physical theory is to predict the behavior of the members of a family.

In order to create a theory, we use methods of objective logic, mathematics and planned experiments, which do not exist in nature. Logic has to serve us with objective rules of justification. These rules should be in compliance with accepted general laws of physics. In this manner, a connection between mathematical theories and physical reality can be established. Theory is, indeed, our construction, and we have to prove its validity. Although we do not offer "the absolute standard of rationality" while creating a theory, we do assign more important and active role to logic in physics. The conclusion, at which I arrived, in the course of the development of the new theory of elasticity, is that logical analysis must be a part of the construction, testing and proof of every physical theory.

In 20th century physics it was a trend to treat logic as a subject almost identical to mathematics. The definition of logic as a subject, which has its own rules of operation that are common to all sciences, is too vague. Logic needs to be distinguished from mathematics for these subjects to have different purposes and methods. The main purpose of logic is to justify a theory. Mathematics has no such purpose or method. Logic has two methods of justification, i.e. inductive and deductive.

The inductive method is used for the justification of selecting the variable concepts and logical structure of a theory. The inductive method in this work is defined as the justification of the particulars of a theory, which are under the umbrella of accepted general definitions and laws. Thus, the law of conservation of mass-energy refers to all particular cases for every closed whole system. All that is considered to be true in general can be applied to relating particulars. However, we cannot say that what we observe in a particular case can be applied to seemingly similar objects with the same measure of certainty. Our observation of repetitious behavior of a particular object doesn't lead to a generalization of behavior for similar objects. It is necessary to find regularity and build a theory for a class of different objects in order to make an inductive justification for the theory. Only a law that is applicable to a class, or classes, of different objects allows us to make inductive justification. In fact, the essence of general laws is that we cannot find an exclusion from such laws. The inductive method is a necessary part of logical justification in the process of constructing and testing a physical theory. For a class of structures, if we found that change of each object is proportional to the force causing the change, and inverse to the inborn inertia of each body that is resistant to change, then general law can be applied to different classes of objects in respect to the change. We may call it a law of change in physics.

The deductive method is another method of logic to employ while structuring a physical theory. In classical logical argument there exists a transition from proposition to conclusion. The assertion is made that if a proposition is correct, then the conclusion should also be correct. The same statement in the concise form is written: "If P, then Q". Actually, this argument needs a correction. Proving a proposition is impossible without proving the inference as true. The logical demand to have an inference requires the certain mathematical structure of a theory, namely, it requires a non-linear propositional physical function. In mathematics no meaningful inference can be made from a linear function. Such a function is logically insufficient. On the other hand, a mathematical differential inference can be made from a non-linear function. The function of a change, which describes a class of physical objects, is a non-linear physical function; change of a whole object is proportional to the internal force corresponding to change, and inversely proportional to the resistance of an object to change. The complete description of the law of change should include a propositional non-linear basic equation as well as the derivative conclusion from this equation. A derivative equation describes the rate of change of a function respectively to the change of the independent variable that is inertia or resistance. The innate property of resistance varies for every member in a class of objects. Thus, a nonlinear function of change allows us to make a comparative analysis of behavior of the members in a family. Without comparison, there is no understanding of change. Although general physical laws are our assumptions, in this methodology, we consider these laws as objective knowledge. The reasons for that are: first, we need the common laws for our scientific models of the universe, and, secondly, universal laws don't contradict our observations. The law of conservation of energy and momentum in a closed system is the foundation for detecting changes. But, in order to predict changes, we need a theory of change.

The logical structure of a physical theory can be developed only by the conjunction of both methods of justification. A theory should comply with general laws in physics, and it also has to be proved. Proof is impossible without inferential certainty. For instance, the propositional equation of change states that change is proportional to a force corresponding to a change, and inverse to the resistance of an object to the change, $\Delta = F/kR$. The elastic force 'F', which is required for a particular change and cannot be measured directly, can be established using inductive justification from the general law of conservation of energy and, specifically, from the calculation of work that is, or would be performed, on a body. The propositional equation by itself cannot be proven because, as of yet, it remains an incomplete description of the function of change. Every variable in the conceptual basic physical equation can be tested experimentally, directly or indirectly, and supported in particular cases by the facts. This is because the variables in a propositional function and its result have absolute values. Then, what other proof do we need for a theory? The purpose of a physical theory is to predict the behavior of objects belonging to a class. However, no one prediction can be made for all members of a given class based on an experiment conducted on a singular member of the class. Experiments can only provide data for concrete occurrences. Instead, we need a law.

The law of change should give a unique result for every member of a class. This means that only one concrete description of the law can be correct. The same result of change for a singular member can be obtained with different equations, by substituting variables in an equation with their descriptions. For instance, in the theory of elasticity description of resistance of a structure to deformation can be written as $R=kA/L$, where resistance is proportional to the area of a cross-section, inversely proportional to the length of a structure, and 'k' is a coefficient of specifics. After such a substitution, the result for deformation $\Delta=FL/EkA$ will be the same as in the original equation, $\Delta=F/ER$, but the relations among the variables in the equation are changed, the character of the function is changed, and partial derivatives from these functions are different: In one of those equations $\Delta=F/ER$ independent variable is 'R'; after substitution the independent variable is 'A'. In one case $d\Delta/dR=-F/ER^2=-FL^2/EkA^2$ and for independent variable 'A' $d\Delta/dA=-FL/EkA^2$. We would be unable to predict behavior without deciding which of these functions is correct because derivative equations describing behavior of functions are different. Logical analysis helps us select the proper propositional description.

Logical analysis includes a system of equations. In such a system, if the derivative inference from a basic propositional equation of change is correct, and the basic equation is supported experimentally, only then is this system of equations justified. The derivative equation contains a relative value as its result, $d\Delta/dR = -F/kR^2$. Relative value $d\Delta/dR$ can be calculated, but not measured. The derivative is correct if the independent variable is selected correctly. A solution can be proved by identifying a derivative with a universal trigonometric function. The derivative equation, which belongs to the family of tangent function, provides certainty to the rate of change for a propositional function that is supported experimentally. Such a derivative function has values that correspond to the tangent function, and can also be proven experimentally in some cases, i.e. in the interval of rapid changes. In this manner, we can build a reliable predictive system of equations for a family of similar objects.

The logic of physics and logic in mathematics display similarities, as well as differences. For example, any non-linear mathematical function and its inference are correct, when presented as abstract statements. This approach is unsuitable in physics. In physics we are concerned with the meaning of function, which describes concrete physical relations. In physics only one description among a number of possible mathematical descriptions is correct. The physical content of a mathematical function places restrictions on mathematical transformations. The rules of logic that can be applied to physical functions and procedures allow us to reach necessary exclusiveness by justifying physical functions and mathematical procedures. Thus, mathematics considers inductive inference as a transition from "next to next" in the order of natural numbers. In physics a definitive inductive transition can only be made from general laws to specific or partial conclusions and laws. The definition of inductive inference, as a logical proceeding from general to specific, can be employed by logic in general, for it seems applicable to all physical sciences. Mathematics has its own constructions, such as a system of basic and derivative functions, which is useful in deductive inference. From the point of view of the new non-linear theory of elasticity, one may see that it is impossible to overestimate the necessity of a logical system in a physical theory.

The logical system of functions, which leads from the selected initial data to the singular definitive description of relations, is necessary for any physical theory. It is impossible to prove an individual mathematical statement as physically acceptable, but it is possible to prove a system of statements based on how they correspond with each other. It is necessary to have a logical-mathematical structure that may connect a complete physical theory with physical observations. Classical logic and epistemology reject the possibility of such a logical structure that can be proven to be true to the facts. However, as described in my works, the logical structure of the Theory of Change provides certainty for its predictions. The *Non-linear Theory of Elasticity* describes a structure that is consistent in all its parts. I call such a logical system definitive logic. Thus, the role of logic in science can be seen when applying rules of inductive and deductive inferences, justifying mathematical structure through definitive logic, justifying terms and conclusions of a theory.

In physics the change of a whole is connected with its properties. A propositional statement is hypothetical until it is proven experimentally and through means of logic. In order to be proved or disproved, certain assumptions need to be tested. According to classical logic, it is necessary to make a transition from a proposition to an inference, and to test the inference. If the inference is correct, then we may conclude that our proposition is also correct. Transitioning from a propositional statement to a conclusion in logic is known as 'argument'. Let's consider general arguments in different logical systems. Classical logic has the following well-known argument in the foundation of its reasoning: "If these facts are admitted, then this conclusion must be admitted," or more concisely and formally, "If P, then Q." In modern deductive logic the definition of a valid deductive argument is that "an argument is defined as a valid deductive argument if its premises could not all be true while its conclusion was false."(Robert J. Ackerman, Modern Deductive Logic) The difference between the classical argument and this modern variation is that the argument, "If P, then Q," has direction. It is asymmetrical. (From 'P' inferring 'Q') The argument in modern deductive logic is an attempt to correct a shortcoming of the classical argument "If P, then Q", which doesn't necessarily lead to a proof. At the same time, by negating conclusive deductive construction, the latter argument is telling us that we cannot prove a conclusion. We can only disprove it. The positive and incomplete argument of classical logic, versus the negative approach to the argument in modern deductive logic, does not provide us with the means to build a successful logical structure. In all these definitions of the validity of arguments, too many assumptions are made. Thus, it is assumed that premises and conclusions can be verified separately from each other as either true or false. It is also assumed that a conclusion can be transferred to a connected statement, and that the empirical judgment provides sufficient means for justification, i.e. the means for testing a proposition and conclusion are an empirical validation of statements. It is often assumed that if the terms of a propositional function are experimentally obtained values, and if the propositional function has an experimentally confirmed value, then not only is the proposition true, but the properly obtained mathematical inference also must be true. From a mathematical standpoint, this would be the case. But this is not the case in physics. Empirical support of a proposition is not logical proof, and it is also not the proof of a conclusion. The fact of the matter is that empirical means are not the only means available to validate or falsify a physical theory. Not all terms and statements of physical theory can be tested empirically. For example, it was shown that the mathematical inference of a propositional function, in principle, couldn't be measured and tested empirically, for the inference represents the logical-mathematical relation of function and independent variable. Relative values in physics, principally, cannot be tested empirically. They need to be tested through means of logic. Physical theory needs a logical structure in its foundation in order to be validated.

Summary: This chapter discusses problems in physics that can be solved using rules of logic. We know that if change in an object occurs then there should be a reason or cause for it. Cause is always another object that comes in contact with an object of interest. Also, cause can be considered to be the energy that changes an object. But, can cause logically be connected with effect within the same equation? Observations and analysis may provide us with clues that would help separate cause from effect. However, there is no functional connection that can be established between cause and effect, for cause and effect belong to independent objects. Different objects in contact may exchange energy. Forces, which arise in contact, are equal in value and opposite in direction. In the struggle for the same space between objects, a force brings changes to each of the objects in contact. Because objects can be different, there is no predictable change that only depends on cause. However, we may consider change in a series of objects, and come up with an equation describing the changes. Change can be presented as a function of reactive force, and characteristics related to change, in an object. Change is proportional to force, and inversely proportional to the resistance of an object to change.For constructing a reliable theory, there is a need in the logical rules of construction propositional statement that can be justified and used in the theory. There are no such rules in classical logic. One of the logical demands here is to construct a propositional statement that can be justified. In order to test such statement, it is necessary to make an inference from it, and justify the inference. One can describe change with a hypothetical physical equation following some experiment. The result of an experiment, or results of similar experiments, will be repeated as well. That doesn't mean that such an equation describes relations correctly. The equation still remains hypothetical. Many descriptions of the same experiment can be written with different variable concepts that would have the same results corresponding to the facts. Mathematics allows us to do this by substituting a variable, or variables, with equivalent descriptions. However, physical conclusions from those seemingly equivalent descriptions appear to be different for the same experiment. It is obvious that for a reliable prediction we need to know how to make an adequate assumption. There are rules in calculus which allow us to obtain a derivative of a function. Mathematical rules do not allow us to solve the problem of selecting variables, and, in particular, mathematics does not allow us to view an independent variable as a single concept, properly.When testing an assumption, classical logic suggests we test an inference, and if the inference is supported by the facts, then the assumption is also correct. In order to have a meaningful inference from a physical theory, its basic propositional equation should be nonlinear. If the empirical test is the only means that is needed to support this logical structure, then classical logic seems to fulfill its role by constructing linear theory. However, logical analysis shows that a linear equation is an incomplete and inconclusive description. One may see the logical insufficiency of such a construction in the inconclusive character of the propositional statement for a single object. Thus, in the theory of elasticity a number of descriptions can be made using the same initial data to obtain the same result. This doesn't mean that we deal with equivalent descriptions. Classical and modern logic has no rule that would require the logical selection and confirmation of a propositional statement. However, if the main argument of classical logic is understood as a transition, not beginning from the data all the way to the conclusion, but starting from a propositional statement and finishing at a concluding statement, then the argument is a logical structure. Classical logic has no specific rules for creating logical structure. Empirical validation is considered the only means of testing a physical theory. But, not all the terms of a physical equation can be tested experimentally. The non-observable terms need logical justification. The logic of relations, principally, cannot be justified empirically. We can measure observable characteristics. But, these characteristics are not necessarily proper concepts in a function. Besides, a function has non-observable concepts as well. An experimental test may validate quantitative relations of a function, but not the qualitative relations. Even if we infer a mathematical conclusion from a propositional function, this new function, from the original hypothetical function, would also be a hypothetical function. Making an inference is a necessary logical step. This step is usually missed in physical theories. It is missed in any linear theory. A derivative function is a logical conclusion, and in its totality, cannot be tested experimentally. It can only be tested in the intervals where rapid changes occur in a propositional function. The universal derivative function in a family of similar objects, which is a smooth continuous curve, is usually a tangent function. A system consisting of a proposition and inference is justified if the propositional function can be supported experimentally, and the corresponding inferential equation is a universal mathematical function.

The Theory of Change

"Ideas are operational in that they instigate and direct further operations of observation; they are proposals and plans for acting upon existing conditions to bring new facts to light and to organize all the selected facts into a coherent whole." (Logic, John Dewey)

In physics we gain knowledge about structures of the universe and it parts. Some knowledge can be obtained by way of empirical means: observation, experimentation and measurement. The physical universe is also a dynamic place. Changes of objects are constantly occurring. In physics we analyze the physical changes and processes in objects and systems. In order for an analysis to be successful, the object of research needs to be considered as an independent whole with internal forces in equilibrium. When an unbalanced external force is applied to an object it changes the energy of the object, its geometry and the position of the object relative to the other objects and/or to its previous position. In order to predict changes it is necessary to have knowledge of the general theory of change. The theory of change describes relations among selected properties of a whole related to the overall change of geometrical characteristics of the whole.

Changes occur due to work that is performed on an object. By definition, in classical mechanics, work is a product of force and the change of geometry, or position, of an object in response to an unbalanced force. If change of an object is observed, then a force was applied and work was performed. The concept of force is a necessary part of understanding changes in the physical world. If there is no change of an object, or no change in its habitual perpetual movement, then no unbalanced force is involved, and thus, can not be detected. The concept of force is associated with the change of a whole. The reactive or inertial force is one of the variables in the function of change. Other concepts in the theory of change relate to the geometry of an object and/or its relative position in space.

A theory is a description that should follow and obey certain rules. The methods that physicists use for studying the properties of physical objects, such as observations, experiments, and measurements are mainly empirical. On the other hand, besides empirical methods, building a theory requires sophisticated methods, which are our intellectual inventions, such as logic, mathematics, and already formulated universal laws and physical definitions. In physical theory we do not describe relations between the parts of a whole object, but rather the relations among characteristics of a whole, which are related to change. In order to predict the behavior of different objects it is necessary to build a consistent theory that can be proven. Physical theory has concepts that have physical meanings. We need rules for selecting concepts related to change. Relations among selected concepts should be provable, via the design of a theory, and by supporting the theory factually.

There are the laws that are considered fundamental in a physical system of knowledge. A physical theory should not contradict universal laws in any way. Also, there are mathematical functions that are true by way of their design, and can be used in building physical theories. Theories are built upon having priori knowledge that is in agreement. We study objects, structures and processes in the universe not only for satisfying our curiosity, but, also, with purposes to predict, use and control processes or changes. Our purposes are better served with the help of theories; and we should construct useful theories.

The progress in the design of theories is connected with the development of mathematics and, in particular, the invention of Calculus. Isaac Newton, who invented the methods of Calculus, considered mathematics to be a common philosophical foundation of physics.

"Since the ancients made great account of the science of mechanics in investigation of natural things; and the moderns, laying aside substantial forms and occult qualities, have endeavored to subject the phenomena of nature to the laws of mathematics, I have in this treatise cultivated mathematics so far as it regards philosophy" (Preface to the Philosophiae Naturalis Principia Mathematica, by Sir Isaac Newton).

This foundation of mechanics put aside "occult quantities" in physics, and at the same time, "laid aside substantial forms," even though forms are substantial for understanding the concept of a whole in nature and the concept of change. Identifying the philosophy of calculus with the philosophy of mechanics also creates flaws in mechanics. Although geometrical properties, such as the length of an object, can be obtained through the summation of corresponding infinitely small parts, the physical properties of a whole cannot be treated this way. Physical properties of wholes are not the properties of their parts.

The new direction of research in mechanics has a new philosophy. Physics needs a unifying plan for building and proving physical theories. There is still doubt that it is possible to prove a physical theory. The common expression is that "it is possible to disprove a theory, but it is impossible to prove it." This opinion exists, partly, because no distinction has been made between studying the order in the universe and establishing order in theories. The first question on this road of building common objective methodology of physical theories is epistemological, namely, how a theory corresponds to studies of nature. Considering already existing theories, we may say that the studies of objects and structures of the universe, in most cases, have a small bearing on the theories that describe changes. The methods of studying entities in the universe differ from describing and predicting possible changes of entities. While studying process/change we select the characteristics of wholes that are connected to the process/change.

Concepts in physical theory are not necessarily obvious characteristics of an object. The concepts are, rather, related to our view of a process. The characteristics of a whole related to a particular process become the terms of a mathematical system of a theory. The methods of studying objects and their structures and parts are mainly empirical: observations, experiments and measurements. Building theories involves empirical, logical and mathematical methods. There is no need to prove empirical data. Nature has no need for the justification of empirical data, and doesn't have logic for us to follow or investigate. On the other hand, we are the creators of physical theories. We have to prove our ideas and schemes. In a theory we not only need a rational analysis of a phenomenon, but a rational structure of a theory, as well. We need a logical scheme for the foundation of a theory. The terms of a theory have to be selected in a way that these terms can be supported empirically or theoretically. Theoretical terms have their roots in a priori accepted knowledge.

A theory should be in compliance with a priori knowledge:

(1) One of the fundamental laws in physics is the law of the conservation of mass-energy. The law states that the total energy and mass in a closed system remains constant unless an outside force or energy acting on the system changes balance in the system. Every theory should be in compliance with the law of mass-energy conservation.

(2) Also, for a priori knowledge we must allocate universal mathematical functions. The function of tangent, for example, plays an important role in constructing and testing a physical theory, and thus, predicting physical changes. This mathematical function is a necessary instrument of analysis for physical theories.

(3) Usually, in physical theories we consider the changes that occur in wholes. In physics a whole is matter-energy organized in a particular space. Besides the obvious ordinary objects, which we consider to be wholes, there are also electro-magnetic energy formations that can be considered wholes, as well.

(4) Similar to matter, organized space has specific properties. Studies of the properties of wholes should also include studies of the properties of space and the properties of matter in the whole. Contemporary physics adheres to the idea that knowledge of the elemental particles and forces will give us knowledge of any and all objects and processes in the universe. This is, however, a metaphysical idea that has no physical foundation or proof. The idea does not take into consideration the properties and energy in the space of a whole. The space of a whole is larger than the sum of the spaces occupied with its parts, unless we

consider a whole as one of its parts. In general, a whole has a different geometry from the geometry of its parts situated at different levels throughout the whole. The properties of a whole, such as elasticity, strength and distribution of forces, depend on the geometry of the whole, as well as the characteristics of matter constituting the whole.

(5) A physical whole is an organized and unified entity of space-matter. A whole consists of different parts. A whole and its parts both have geometrical structure. Space is occupied with matter at different levels of organization. Subatomic elemental particles, nuclei, atoms, atomic-molecular structures, macro-structures, ordinary-sized structures such as geometrical wholes, systems of structures, celestial bodies and structures are all entities that can be studied as independent wholes. These parts of a larger whole can be considered as independent wholes themselves, for they can be separated and studied independently. Moreover, each whole is a part of a larger whole. Each of these part-wholes has its own properties. In this hierarchy of parts, the properties of a whole cannot be inferred solely from the knowledge of properties of elemental particles and structures.

(6) The properties of a whole, in general, are not the properties of its parts. The entity's autonomy is the reason for this. The characteristics of a whole that is related to its parts have empirical statistical values. There is no functional relationship between independent entities situated at different energy levels of space. Also, there is no functional relationship between a whole and its parts.

(7) Force acting on a body is associated with another body. Conjunction of the bodies causes equal, but opposite, forces to be applied to these bodies. Force is a theoretical concept in any theory. The value of such a force cannot be

measured, rather, it can only be found through other measurable concepts with equations of equivalence, and through common definitions. The geometrical parameters of an object, on the other hand, are measurable. For the construction of a theory, it is important to use the duality of space-matter, which allows for the construction of a function by connecting the concept of force with a change of geometry, or with a change of the relative position of an object.

(8) Physically, the matter and space of a whole are non-separable. In physical theories, however, we include the properties of matter and space separately. We tore apart the unity of space-matter for the purpose of revealing the relations between space and matter. Only in such unity can the change of one characteristic of a whole correspond to the change of another of its characteristics. Thus, we can find and describe functional relations between inseparable parts of a whole. If a pattern is found, then that would allow us to make a predictive quantitative description for a family of similar objects.

(9) The proper logical structure of a theory makes it possible to select the physical terms of a theory. The terms can be observable, as well as theoretical non-observable. Theoretical terms usually come from the priori established definitions in physics. Thus, the understanding of the term "force" may come from the definition of work. Work needs to be performed to change geometry and/or the position of an object. Work can be calculated. It linked to the law of conservation of energy. Mechanical work, by definition, is the product of force and change of the relative position of an object. External force is identified with an independent object-cause that pushes or pulls an object under investigation.

(10) Work and the corresponding concept of force are theoretical physical concepts. Work is equal to the energy required by, or spent on, change. When we observe changes we attribute them to a force acting on an object. This happens to be the case even when we don't see an object causing a force, as in the case of elastic force, or in the cases of changes caused by the presence of an invisible celestial body/energy in a celestial system. Force is a theoretical physical concept, which is included in the definition of a physically measurable process. In the theory of elasticity, the elastic force in a small volume of a whole, or in other words, stress "σ", is proportional to the deformation, or strain "ε", in this volume. Deformation is an observable concept that can be measured. Invisible stress or force can be calculated correspondingly, $\sigma = E\varepsilon$ (Hooke's law). Experimentally obtained the statistical coefficient of proportionality "E" depends on the material of the whole.

(11) In all cases, a physical event is the result of contact between two or more independent entities. One of the entities in contact plays the role of cause, and the other, effect. During a scientific investigation of an event in a theory we have to separate cause from effect, for there is no functional connection between them. We deal with two independent objects, while functional relations exist only for the properties of an entity, which are intimately connected. There are no inherent properties that designate one object to be a physical cause, and another object the effect. We may switch sides and consider what happened to the object that was previously the cause in this conjunction. There is no consecutive relationship between cause and effect. The idea that first appears as a cause, followed by some effect, is a metaphysical idea. The conjunction of

physical objects appears simultaneously in a cause-effect relationship. The roles of cause and effect are assigned in the process of scientific analysis. We separate cause from effect for the purpose of analysis and predicting of a phenomenon. Without such separation it would be impossible to come up with a proper mathematical description of a phenomenon.

(12) All means we use for creating physical theories should be adjusted to the physical purposes. This also refers to mathematics. The physical nature of physical theories calls for special rules of mathematics. In mathematics symbols and their relations in equations are considered to be abstract. In physics the relations among the terms are specific rather than abstract. In physical theories we have to select proper terms that cannot be freely replaced by seemingly equivalent mathematical expressions because after replacement the relations between terms are changed. The rules of transformation of mathematical functions have to be checked for appropriateness in physics.

Hypotheses and Facts

H1. Physical theory can be falsified, but it cannot be proved.

A theory can be falsified if it contradicts the physical facts, or if it doesn't obey the universal laws of physics.

A theory can be proved within the universal laws of physics. In order to do this, a theory should have a consistent logical structure; the physical concepts of a theory should fit/satisfy this structure, and the relations between the concepts need to be proved by the design of the logical structure/scheme and the empirical tests of the measurable concepts.

H2. Physical theory reflects order in nature.

Physical theory is not a reflection of order in nature. The physical world consists of matter, space and energy. This building material is organized in the structures of space-matter-energy. On the other hand, a theory is a mathematical description of relations existing between selected characteristics of a physical whole. Said relations can be revealed in the process of the change of a whole. We build physical theories according to some logical, mathematical and empirical rules, and as a result can prove them, but at the same time, cannot prove our models of universal structures.

H3. A theory is valid if it has the experimental support.

"Theory can never be proven, because nothing in theory has demonstrable physical nature that can be isolated and examined. All that can be proved is that theory fits empirical data," (Empirical evidence and theory, James A. Putnam)

There is no experiment that may tell us that a hypothetical proposition is correct or incorrect. According to the logical demand, a proposition is correct if an inference from the proposition is supported by facts. A proposition by itself cannot be validated with experiments, for it is only a part of the description of the relations among the selected concepts. The inference from a proposition has a relative mathematical value as its result. The relative value, in general, cannot be tested experimentally. The only way to verify a theory is to consider a system of the propositional equation and corresponding derivative equation. Physics has many unproved theories because of that erroneous belief that if a mathematical expression is created in a way that meticulously fits the experiment, then a theory is satisfactory. From the methodological point of view, such a theory is not a reliable theory, yet. Proof of a theory requires not only empirical validation, but the logical validation, as well.

H4. The properties of a whole can be determined if the properties of the elemental particles and forces are known.

"By and large, a system in classical physics can be analyzed into parts, whose states and properties determine those of the whole they compose." (Stanford Encyclopedia of Philosophy)
A whole consists of parts or a hierarchy of parts. However, not only are the properties of the parts are not the

properties of the whole, but also, the properties of a whole are not the properties of its parts. The properties of a whole cannot be deduced even with a complete knowledge of its parts and universal forces acting on the whole. In some experiments, the properties of the parts could demonstrate themselves. For understanding the behavior of a whole, it is necessary to know the properties of the parts, as well as the properties of the whole.

H5. The properties of a whole are the properties of its matter.

A whole consists of matter that is organized in the space of the whole. In reality, we cannot separate matter from space, but in a theory we can do this. We may compare the statistical properties of a whole depending on matter with the properties depending on the space that this matter occupies. The property of a whole demonstrated itself in an experiment can be of different nature. It can be a property of matter or it can be a property depending on space of a whole.

H6. Light has a double nature.

The organized structure of light can be considered to be a whole. In different experiments, a whole may demonstrate the properties of a complex geometrical whole, or the properties of it parts. The property of light, in some experiments, may belong to light as a whole structure, or in other experiments, it can be the properties of the constituting parts of light. This observation concerns not only light, but the solid bodies, as well.

H7. We can predict the behavior of a whole if properties of parts are known.

We cannot predict behavior of a single whole even with complete knowledge of the properties of it parts and the properties of a whole.

The behavior of a whole can be understood and predicted only in comparison to the changes in a family of similar wholes.

H8. We can select the terms of a physical function following an observation.

. The terms of a theory are not necessarily the observed characteristics, but often, the terms are the complex physical concepts related to a change.

In order to prove a theory we need a logical structure of the theory. The terms of a theory should be selected to fit such a logical structure

H9. Internal forces acting in a whole are keeping its matter together.

If a whole is not in the process of change, then no forces that keep the whole as it is formed can be detected. No unbalanced force can be registered unless the unbalanced force is applied to a whole. A force is needed to create a whole from several parts, destroy the whole, or force induces change of a whole.

H10. The relationship between physics and mathematics is harmonious.

The appearance of physical ideas in mathematical forms overshadows real relationships between physics and mathematics. There is no unity born from some hypothesis. A physical equation distinguishes itself from a pure mathematical equation with its essence. The terms in mathematics are abstract. The terms in physics are concrete, and the relations between terms are unique. Mathematics is a scientific tool in physics, and should serve the purposes of physics. Not all mathematical rules are acceptable in physics. Thus, the most general rule of mathematics, that the function of a function can be presented as a single function, doesn't serve physical purposes well. The substitution of a variable concept with an expression, which represents other concepts, changes relations in a new equation, and this may change an inference that is a physical conclusion, as well. Physical theory cannot be proved unless we select variable-concepts properly, according to logical rules.

My Journey to the New Methodology

As a mechanical engineer I was interested in solving practical engineering problems, such as establishing safe and reliable dimensions of structures, and predicting the behavior of a structure. This appears to be an impossible task without the knowledge of the individual limiting stress in a structure. But, there is no method for calculating the individual limit of a structure in the classical theory of elasticity. Another practical problem: Manufacturing products usually come in a series of similar structures. And in the classical theory of elasticity not a single criterion of similarity exists between the designs of any two structures, which can project success. Every particular model in a series has to be destroyed in order to find an approximate individual limit. There is no method for the optimization of dimensions of structures in classical theory. These are the problems in the field of engineering, and no sound theory exists to solve them. Many engineers, including myself, have attempted to solve everyday engineering problems with the help of classical theory and statistical methods that are based on engineering practices. However, soon I realized that it is an impossible task to solve these problems without a revision of the fundamentals of engineering design.

The phenomenon of elasticity has been discovered by the English physicist Robert Hooke in 1660, and the article published in 1678. In his article Hooke said that all materials and all structures have elastic properties. Elastic deformation of a structure caused by an external force disappears after this force is removed. The elasticity of a body has a limit. It is important to know this limit for building safe and reliable structures. Adequate theory is essential for solving the above mentioned engineering problems. I have been thinking over the ideas of comparative analysis of structures and optimization since 1968. Only in 1986 was I able to test my method with a couple of experiments. The experiments and previous comparative analysis led me to the new point of view on the nature of the limit of elasticity: The individual limit of elasticity of a structure has a relative nature. It can be the limit of material, or a limit imposed by the geometry of a structure, depending on which one has lesser value. Those limits are not connected functionally. Also, for the comparative analysis of structures, the basic mathematical formulation of the law of elasticity has to be changed. The newly formulated law should state that elastic deformation is proportional to elastic force distributed in a structure, and inversely proportional to resistance of a structure to deformation. In order to prove this theory I had to revise the principles in approaching physical problems, include logical scheme in my theory, consider the demands of physics on mathematics, and consider the roles of empirical methods in physics.

There was no developed methodology in physics that required logical structure. The necessity of proving a theory is obvious, but it is only possible if its methodology includes the demand of proof. Analysis of the sources of inadequacy of the Linear Theory of Elasticity led to the justification of the Non-linear Theory of Elasticity. Following, is the reasoning for establishing the main general principles of methodology in physics.

(1) Physical relations have a limit. Therefore, these relations require a description with a non-linear function. This demand has several physical and logical reasons. Elastic relations have limits. Although the deformation,

elastic force, geometric form and material relations have a limit of elasticity of material as a boundary condition, these relations also have an observable limit that depends on the geometry of a structure. This structure-specific limit usually is the actual limit of a structure. A continuous function describing deformation, elastic force, geometric and material relations, includes the limiting elastic deformation, be it the limit of a material, or a specific limit generated by the geometry of a structure. It is necessary to have a mathematical method to detect the limiting deformation.

The limit of elasticity is characterized by a rapid increase in the rate of change of deformation. The Linear Theory of Elasticity operates with the linear function of stresses and deformations. Such a function does not differentiate the change of deformation adequately for identifying the limit. A rate of change for the linear function is constant. The practical demand of finding the limit of elasticity of a structure leads to the conclusion that we need a non-linear description of elastic deformation. The mathematical means should be adequate for serving physical purposes.

(2) The physical nature of the Theory of Elasticity, and its practical purposes, require verification of their mathematical descriptions. Verification, according to current norms of logic, refers not so much to the verification of a proposition, but rather to the confirmation that the inference from the initial hypothetical description is true to the facts. A mathematical inference from a basic function is a derivative function. In order to conduct such a test, we should have a logical-mathematical system consisting of a basic equation and its derivative equation. One of the disadvantages of the Linear Theory of Elasticity

is that a linear basic function cannot produce a meaningful derivative function. We have no way to test linear theory. The logical demand of the verification of a theory requires a non-linear description of the basic propositional equation.

(3) Mathematical descriptions of physical theories are linked, as a rule, to our hypothetical physical ideas about phenomena, rather than to logic. The physical idea in the foundation of the Linear Theory of Elasticity is that deformations are so small in comparison with the dimensions of a structure that the relation between them can be represented by a linear function. However, the concepts "small" and "large" are relative. Elastic failure occurs within a range of small deformations. In order to see and appreciate changes of deformation, an analysis must be made at the level of the values of deformation, rather than at the level of the dimensions of a structure. Mathematically, such analysis is possible if the selected independent variable relates to deformation describing an effect of geometry of a structure on deformations. A new concept of geometrical stiffness may correspond to the requirement of compatibility to change of deformation with stiffness values.

(4) Elastic relations are physically definitive, as should be their descriptions. The definitive description includes basic and derivative functions. A basic equation gives the absolute values of change corresponding to a certain individual value of the independent variable concept. The derivative equation gives the relative position of each point in the continuous function, and thus, describes the elastic behavior of an entity depending on this position. The drawing illustrates the point of view that a description of physical relations should be concrete. Two curves are going through the point A(x, y). If a real structure belongs

to curve #1, its relative position is characterized with a slow rate of deformation, dy/dx=-tanα, and the structure is in position of elastic stability. If point A(x, y) belongs to function #2, then it is in a position of elastic failure where small changes of the variables produce a rapid increase of deformations. In the real physical world only one behavior is true for a given structure, and thus only one description is correct. In order to find which of the possible descriptions is correct it is necessary to examine the equations at both levels, for the distinction is revealed in the derivative equation.

(5) Among a number of descriptions having the same variable concept, such as Y=f(X) and Y=f (F(X)), only one description can be physically correct. In mathematics, the function of a function can be presented as a single function. In physics such a transformation can be the cause of a fundamental misconception. The functions obtained in such a manner are not physically identical. Although in both functions the same value of 'X' produces the same values of 'Y', these equations have different derivatives, depending on whether 'X' is an independent variable, or whether it is a variable part of another independent variable concept, as this selection implies different physical behavior. Physical determinism requires the selection of one correct description, while excluding other descriptions. The double-level, logical-mathematical system of equations in physical theory needs verification.

(6) In classical logic, verification is the process of corroboration an inference from an initial hypothetical description by using corresponding facts. But, theoretically and practically confirmation of a physical theory cannot be achieved with the empirical verification of an inference. We deal with two types of equations. The basic equation

operates on the absolute values of the variables, Y=f(X). The derivative equation, on the other hand, has a relative value as its result, dY/dX=f '(X). Absolute values can be measured, and a basic equation can be validated experimentally. However, results of derivative equations are relative values. They can be calculated, but not measured. The derivative result is correct if we describe the relations in a basic equation correctly. We are in a circle of a logically indeterminate system. There is no experimental proof for the logical structure "If proposition P is correct, then conclusion Q is correct," for it is impossible to measure the relations of the logical structure. Then, for testing a theory we should examine a system of equations, rather than testing an inference.

(7) The objective mathematical tool for testing a hypothetical derivative function can be a universal function that has known properties, such as a tangent function. A tangent function describes possible changes in each point of a continuous function, i.e., small changes in the interval 0<Y<1 and rapid changes in the interval1<Y<10. The system consisting of a basic function and derivative function is correct when the derivative function is a universally known function that does not need verification, and the basic function that corresponds to that derivative can be verified experimentally. In particular, if a continuous non-linear function of deformation occupies the domain of a tangent function, and its derivative at any point is the constituent part of a tangent function, then such a function is uniquely determined. These principles were first laid down in the book *Non-linear Theory of Elasticity and Optimal Design*, by Leah W. Ratner, Elsevier Nov. 2003.

The theory of elasticity is the foundation for the scientific design of machines and structures. It is one of the most charged areas of human activities. The adequate Theory of Elasticity has great practical value. While the new methodology was first worked out in the Theory of Elasticity, this methodology is general for all theories in physics. The new methodology gives a new point of view on the physical universe.

GRAPHICAL DESCRIPTION OF CHANGE

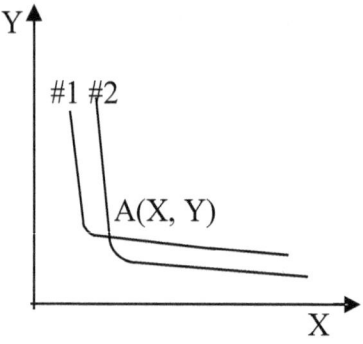

Through each point of curve, Y=f (X) numerous curves can be drawn. All these descriptions give the same resulting change "Y" for a point/object "x=A". But only one of the descriptions can be correct. In curve #1 position of point "A" suggests slow possible changes. In curve #2 for the same point/object "A" it shows rapid changes. Behavior of an object in physical world does not depend on our description. We have to select an independent variable "X" correctly in order to describe adequate behavior and have a possibility to predict it. Tangent to curve indicates rate of change of the function. That suggests way of selection adequate independent variable. Independent variable "X" should satisfy system of basic equation Y=F/X and its derivative dY/dX =tan α.

Leah W. Ratner is a mechanical engineer. She earned her Masters Degree from Moscow Auto-Mechanical Institute in 1956 majoring in the technology of metalworking machines. Since graduation she worked as a senior engineer-technologist; as design engineer; as teacher of "Strength of Materials" in Moscow Machine-Tool College; as senior engineer of the methods of engineering education; as senior research engineer and as senior engineer of technical calculations. Problems in the theory of elasticity and scientific design process came to her full attention in 1969 and since that time she made it her mission to build the reliable theory for the structural design.

Leah Ratner holds two US Patents and the author of the two articles in the Engineering Journals: Machine Design (1986) and Experimental Techniques (1999). She is also the author of the books "Non-linear Theory of Elasticity and Optimal Design" published by Elsevier/Science in 2003 and this book "Order in Nature versus Order in Physical Theory".

Leah Ratner immigrated to the USA in 1979. Presently she is retired and lives in the Great Chicago area.

Book Description

Scientific and technological achievements depend on foundation of the physical sciences. The current affairs in physics characterize the lack of common philosophical platform; there are no logical rules for building theories; no connection between abstract mathematics and concrete demands of physics; and no reliable criteria for testing correspondence of a theory to reality. New methodology that is proposed in this book has been tested in the new theory of elasticity as a reliable foundation of engineering structural design (The book *Non-Linear Theory of Elasticity and Optimal Design*, author L.W. Ratner was published in 2003 by Elsevier).

This book *Order in Nature versus the Order in Physical Theory* concerns new common platform for science and technology.

www.ingramcontent.com/pod-product-compliance
Lightning Source LLC
Chambersburg PA
CBHW051540170526
45165CB00002B/814